# Springer Theses

Recognizing Outstanding Ph.D. Research

For further volumes:
http://www.springer.com/series/8790

## Aims and Scope

The series "Springer Theses" brings together a selection of the very best Ph.D. theses from around the world and across the physical sciences. Nominated and endorsed by two recognized specialists, each published volume has been selected for its scientific excellence and the high impact of its contents for the pertinent field of research. For greater accessibility to non-specialists, the published versions include an extended introduction, as well as a foreword by the student's supervisor explaining the special relevance of the work for the field. As a whole, the series will provide a valuable resource both for newcomers to the research fields described, and for other scientists seeking detailed background information on special questions. Finally, it provides an accredited documentation of the valuable contributions made by today's younger generation of scientists.

## Theses are accepted into the series by invited nomination only and must fulfill all of the following criteria

- They must be written in good English.
- The topic should fall within the confines of Chemistry, Physics, Earth Sciences, Engineering and related interdisciplinary fields such as Materials, Nanoscience, Chemical Engineering, Complex Systems and Biophysics.
- The work reported in the thesis must represent a significant scientific advance.
- If the thesis includes previously published material, permission to reproduce this must be gained from the respective copyright holder.
- They must have been examined and passed during the 12 months prior to nomination.
- Each thesis should include a foreword by the supervisor outlining the significance of its content.
- The theses should have a clearly defined structure including an introduction accessible to scientists not expert in that particular field.

Troy K. Townsend

# Inorganic Metal Oxide Nanocrystal Photocatalysts for Solar Fuel Generation from Water

Doctoral Thesis accepted by
University of California, Davis, USA

Springer

*Author*
Dr. Troy K. Townsend
Department of Chemistry
The U.S. Naval Research Laboratory
and The National Research Council
Washington, DC
USA

*Supervisor*
Prof. Frank Osterloh
Department of Chemistry
University of California
Davis, CA
USA

ISSN 2190-5053 ISSN 2190-5061 (electronic)
ISBN 978-3-319-35883-3 ISBN 978-3-319-05242-7 (eBook)
DOI 10.1007/978-3-319-05242-7
Springer Cham Heidelberg New York Dordrecht London



Printed on acid-free paper

Springer is part of Springer Science+Business Media (www.springer.com)

**Parts of this thesis have been published in the following journal articles:**

**Overall Photocatalytic Water Splitting with NiOx-$SrTiO_3$—A Revised Mechanism.** Troy K. Townsend, Nigel D. Browning, and Frank E. Osterloh, Energy and Environmental Science, 2012, **5**, 9543–9550.

**Nanoscale Strontium Titanate Photocatalysts for Overall Water Splitting.** Troy K. Townsend, Nigel D. Browning, and Frank E. Osterloh, ACS Nano, 2012, **6**, 7420–7426.

**Photocatalytic Water Oxidation with Suspended alpha-$Fe_2O_3$ Particles-Effects of Nanoscaling**, Troy K. Townsend, Erwin M. Sabio, Nigel D. Browning and Frank E. Osterloh, *Energy and Environmental Science*, 2011, **4**, 4270–4275.

**Improved Niobate Nanoscroll Photocatalysts for Partial Water Splitting**, Troy K. Townsend, Erwin M. Sabio, Nigel D. Browning, and Frank E. Osterloh, *ChemSusChem*, 2011, **4** (2), 185–190.

*... to Emily... for her care and support...*

# Supervisor's Foreword

From a conceptual standpoint, the photoelectrocatalytic water splitting reaction is the most simple and least expensive way to convert solar energy into (hydrogen) fuel. Only sunlight, water, and a catalyst are required! But for widespread implementation of this water splitting technology, catalysts must fulfill several criteria: They must have high catalytic activity under solar illumination, they must be chemically stable, and they cannot be based on rare, expensive, or toxic elements. The oxides of the 3d transition elements fulfill all of these conditions, except for one: their visible light-driven photoelectrocatalytic activity is very low. This is a result of several factors, incl. the short lifetimes of photogenerated charge carriers, low charge mobility in the lattice, and high overpotentials for both the electrochemical oxygen and hydrogen evolution reactions. In this Springer Dissertation, Troy Townsend shows that these problems can be addressed by nanoscaling the metal oxide crystals and by chemically interfacing them with metal and metal oxide cocatalysts. Three nanoscale systems were prepared and studied with regard to their nanocrystal morphology, optical properties, energetics, and electro- and photocatalytic activity.

Chapter 2 describes the preparation of well-defined nanoscale composites that employ niobium oxide nanoscrolls as light absorbers, and platinum and iridium oxide nanoparticles as cocatalysts. The photocatalytic activity of these composites is determined by the electrochemical activity of the cocatalysts. This highlights the importance of fast charge transfer kinetics for efficient water photoelectrolysis. Chapter 3 describes the first use of *suspended* hematite ('rust') nanoparticles for photocatalytic water oxidation under visible light. It is found that the activity of the particles is size-dependent and limited by the ability of the photoholes to migrate to the water oxidation sites on the nanoparticle surface. This project complements current research on nanostructured hematite photoelectrodes, which are attractive because of their low cost and high stability. Chapter 4 demonstrates that *overall* water photoelectrolysis can be achieved with a nanoparticulate nickel oxide–strontium titanate–nickel composite. The system extends the known set of functional nanoscale water splitting catalysts to four. Here, the opposite reactivity trend is observed—the larger particles have higher activity, likely due to better charge separation. Lastly, Chap. 5 describes the application of photoelectrochemistry and surface photovoltage spectroscopy to gain insight into the charge transfer properties of the strontium titanate—nickel system. Based on this data, new roles of the nickel oxide as water oxidation cocatalyst and of nickel metal as water reduction catalyst are established.

Overall, these results highlight synthetic and functional aspects of nanostructured metal oxides for solar energy to fuel conversion. By focusing on particle suspensions, not on the more popular photoelectrodes, the work provides a fresh perspective in the emerging field of nanoscale photoelectrolysis. The thesis should be of interest to individuals working in the areas of solar energy conversion, nanomaterials, and electrochemistry.

Davis, December 2013 Prof. Frank Osterloh

# Abstract

Realizing a method for converting sunlight to renewable fuel may be one of the most important challenges of the century. Water splitting to hydrogen and oxygen gaseous fuels can be powered by the sun using light absorbing catalysts. In the pursuit of cheap, active, and stable catalysts, this dissertation focuses on inorganic metal oxide semiconductor nanocrystals for solar water splitting including their synthesis, characterization, and reactivity. The results from this collection of investigations have shed light on the properties of nanomaterials and their light-driven water reduction and oxidation reactions. Chapter 2 shows the effect of cocatalyst deposition onto wide band gap UV-absorbing asymmetric nanoscale niobium oxide crystals. With spatial charge separation in mind, platinum (an electron acceptor and water reducer), and iridium dioxide (a hole scavenger and known water oxidizer) were photodeposited onto $H_4Nb_6O_{17}$ crystals to achieve complete water reduction and incomplete water oxidation. In this case, the water oxidation reaction was shown to be confined to a surface-bound peroxide intermediate. Chapter 3 transitions into narrow band gap visible light absorbing $Fe_2O_3$ nanocrystal water oxidation catalysts. These materials were synthesized, characterized, visualized with high resolution transmission electron microscopy (HRTEM) and tested for their photocatalytic activity under visible light irradiation while in the presence of aqueous $AgNO_3$, a sacrificial reagent. Chapter 4 describes efforts to advance toward a complete nanoscale water splitting catalyst without reliance on sacrificial reagents. In this case, both half reactions proceed and hydrogen and oxygen are evolved from the solution. Nanoscale $SrTiO_3$ (6.5 nm and 30 nm) was loaded with NiOx and properties were compared with bulk $SrTiO_3$ to elucidate the effects of nanoscaling. Chapter 5 outlines the photogenerated electron/hole pathway in NiOx-STO by specifically loading nickel oxide and/or nickel metal cocatalysts onto bulk $SrTiO_3$ crystals. These materials were tested for water splitting from pure water, and surface photovoltage/electrochemical measurements showed evidence of electron trapping in nickel and hole trapping in nickel oxide, giving rise to a three-component complete water splitter.

# Acknowledgments

Without guidance and mentoring from Frank, my journey through this degree would be meaningless. He has challenged me and supported me during my aspirations for numerous fellowship applications, travel awards, and new job prospects. His open door was free for me to approach him with my questions about science and the process of science. Under his mentorship, I have traveled the world to talk about science, published my results in permanent and accessible forms, and grew as a student of general knowledge about how things work. Thanks for your help and continued friendship.

Second, the State of California is not home to me, but seems to be home to my success. The predictable and beautiful weather serves to make things here work. Forward-thinking people trek here to gain high-level educations, and the culture of coastal California is intriguing and enabling. Thanks to California and UC Davis.

Third, thank you to my group members and colleagues who have trained me and supplied me with materials. It's not trivial to learn one's way around a lab or through the over-arching traditions of the chemistry department. Thanks guys.

> A man is relieved and gay when he has put his heart into his work and done his best; but what he has said or done otherwise shall give him no peace—Ralph Waldo Emerson.

# Contents

# Chapter 1
# Introduction

## 1.1 Global Power Demand

The sun delivers 120,000 TW (approx. 1000 $Wm^{-2}$) of power to the surface of the Earth, which far exceeds current and projected human demand. Given the rising population and increasing demand for power, renewable sources of energy must be implemented in order to reduce the detrimental effects of burning fossil fuels. Processing carbon fuels is not renewable and leads to increased pollution and rising global temperatures. Considering the prospect of growing energy usage in developing nations, renewable options that are robust and cost-effective must be realized in order to preserve healthy ecosystems and stable economic growth.

China and India currently contain ~40 % of the world population compared to the USA (<5 %), yet the USA consumes three times more power per person (Fig. 1.1) [1]. In addition, Europeans consume approximately 50 % of the per capita power used by the average person living in the USA.

Making allowance for increased per capita power demand in developing nations, inexpensive carbon fuels are predicted to be the fuel of choice (Fig. 1.2a). Global power demand is projected to rise from its current level (16 TW) to 25 TW by year 2035 [2]. This demand is expected to be supplied by oil (33 %), coal (29 %), natural gas (17 %), renewables (12.5 %) and nuclear (8 %) fuels [2]. In order to reduce dependence on non-renewables, alternative energy sources have been investigated for their merit as a carbon fuel replacement (Fig. 1.2b). Biomass processing is limited by the low efficiency of photosynthesis (<1 %) [3], and geothermal technology has not progressed due to inadequate drilling techniques. Wind energy is limited by low on-shore and near-shore flux, and hydroelectric power is constrained by the number of rivers and streams. In contrast, solar energy harvested by 10 % efficient silicon cells stretched over 0.65 % of the world's terrestrial surface can provide 20 times more power than the projected demand for the year 2035 [4].

Photovoltaic [5] and electrochemical solar cells [5–7] can achieve high quantum conversion efficiencies (55–77 %); however, they remain prohibitively expensive compared to carbon fuels. In order to reach grid parity (point at which

T. K. Townsend, *Inorganic Metal Oxide Nanocrystal Photocatalysts for Solar Fuel Generation from Water*, Springer Theses, DOI: 10.1007/978-3-319-05242-7_1, 

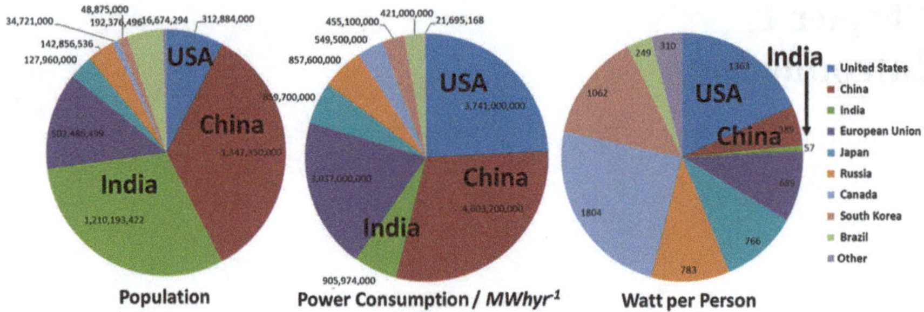


**Fig. 1.1** Global statistics by region: Total population (*left*), Total power consumption (*middle*), and Average per capita power usage (*right*) [1]

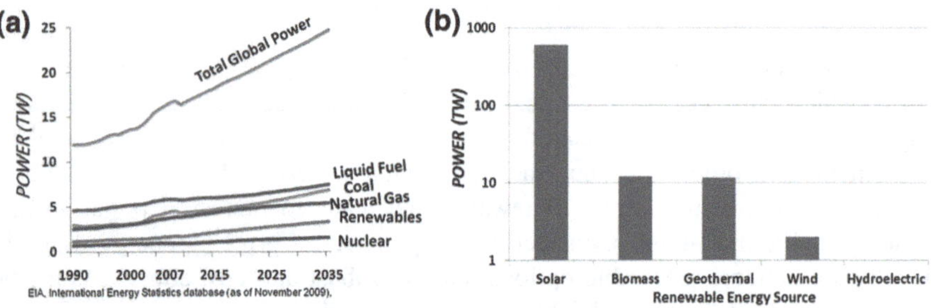


**Fig. 1.2** Predicted global power demand by fuel type [2] (**a**) and log scale of total viable power harvesting potential from renewable energy sources compiled by Nate Lewis [4] (**b**)

the price of solar energy is equal to electricity from the grid), new inexpensive materials and fabrication methods are being explored. One inherent problem with solar cells is the method of daytime energy storage for night usage. Solar cells convert photons to electricity during the day, and the voltage generated can either be stored as a chemical potential in batteries, a physical potential as compressed air or as a chemical fuel from water electrolysis as hydrogen gas. However, batteries are a high-cost and non-renewable route due to generation of chemical waste and the price of ion replacement and waste recycling.

Through a process inspired by photosynthesis, solar water splitting involves a direct conversion of solar energy to storable and portable chemical fuel. Solar cells with sufficient power output can be attached to cathode/anode electrodes to reduce/oxidize water, respectively. However, excessive power conversion losses occur during the light-to-electricity-to-hydrogen steps. In contrast, photocatalysts dispersed in solution convert light energy directly to hydrogen. Semiconductor materials with a suitable band gap and band positions can theoretically act as catalysts for the water splitting reaction to produce storable and renewable chemical fuel from sunlight and water.

Due to the rising interest in nanoscale photocatalytic water splitting as a viable method for energy production, the Department of Energy *Directed Technologies* prepared a cost-benefit analysis of large scale photochemical hydrogen production [8]. Four types of water splitting routes are examined for overall capital costs and final $H_2$ price per kilogram. Given that a suspended overall water splitting nanocatalyst can be improved from the maximum demonstrated efficiency of 2.5 % to a new level of 10 %, which is theoretically achievable. In areas with an average solar insolation of at least 6 kW $h^2$/day, Type 1 (suspended nanocatalysts) is predicted to be the cheapest and easiest option for $H_2$ production: Type 1 = $1.63/kg $H_2$, Type 2 (separate anode and cathode compartments for suspended nanocatalysts) = $3.19/kg $H_2$, Type 3 (submerged solar cells electrically connected to water) = $10.36/kg $H_2$, Type 4 (Type 3 with sunlight concentrator) = $4.05/kg $H_2$. Presently, $H_2$ is collected as a side product of industrial reactions and priced around $4.00/kg $H_2$ Therefore, if the efficiency requirement for Type 1 technology is achieved through research and development, then this could initiate a new generation in renewable energy production.

## 1.2 Photocatalytic Water Splitting

The endothermic conversion of water to hydrogen and oxygen gas can be accomplished with a theoretical minimum energy of 1.23 eV, and the reverse exothermic reaction can release a theoretical maximum of 1.23 eV (Eq. 1.1). In this way, a closed system can perform repeatable redox reactions of water during the storage (hydrogen production) and release of energy per unit time (power generation).

$$2H_2O_{(l)} \rightleftharpoons 2H_{2(g)} + O_{2(g)} \tag{1.1}$$

$$\Delta G = +237 \text{ kJ mol}^{-1}$$

$$(1.23 \text{ eVe}^{-1} \lambda_{\text{min}} = 1100 \text{ nm})$$

Light-absorbing catalysts can lower the activation barrier for water redox reactions in neutral solution so that they occur closer to the theoretical electrochemical potentials of water reduction (Eq. 1.2) and water oxidation (Eq. 1.3).

$$4H^+_{(aq)} + 4e^- \underset{Catalyst}{\overset{hv}{\rightleftharpoons}} 2H_{2(g)} \qquad E^0 = -0.82 \, V \, vs. \, NHE \, at \, pH \, 7 \tag{1.2}$$

$$2H_2O_{(l)} + 4h^+ \underset{Catalyst}{\overset{hv}{\rightleftharpoons}} 4H^+_{(aq)} + O_{2(g)} \qquad E^0 = +0.41V \, vs. \, NHE \, at \, pH \, 7 \tag{1.3}$$

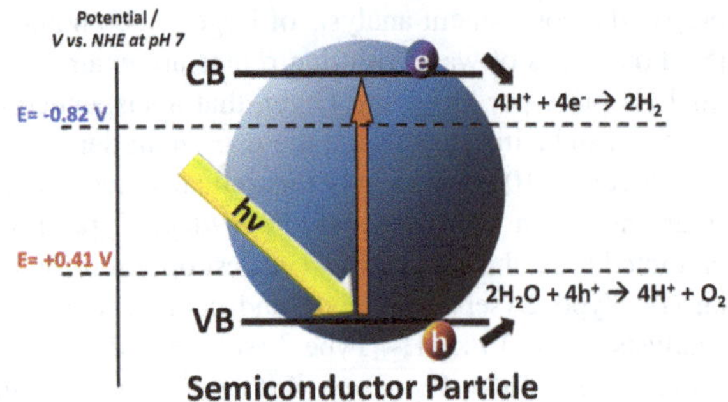


**Fig. 1.3** Photophysics of a photocatalyst nanoparticle suspended in water showing conduction and valence band levels of the semiconductor and theoretical water redox potentials versus NHE at pH 7

First discovered by Fujishima and Honda (1972), light-absorbing semiconductors are known to catalyze these reactions [9]. Their initial study on $TiO_2$ revealed that light ($\lambda \leq 413$ nm, E $\geq$ 3.0 eV) can create an excited electron/hole pair (exciton) on the $TiO_2$ catalyst. Excitons react with water at the semiconductor/liquid interface under an applied electrical or chemical bias potential if the energetics of the catalysts are not favorable. However in theory, electrolysis of water can occur without an applied potential by using a catalyst that satisfies the following requirements (1) the Fermi level of the electron (near the conduction band for n-type materials) must be more reducing than the theoretical potential for water reduction, (2) the Fermi level of the hole (near the valence band) must be more oxidizing than the theoretical potential for water oxidation, and (3) the total energy gap separating the electron/hole pair must be greater than 1.23 eV (Fig. 1.3).

This amount of potential energy is equivalent to infrared light ($\lambda \leq$1000 nm). However, due to overpotentials that arise from electron/hole recombination, high energy intermediates and inherent energy barriers, additional energy ($\sim$0.6 eV per electron/hole pair) is often required. Assuming 2.0 eV contains enough energy to overcome these barriers, active catalysts can operate under visible light irradiation ($\lambda \approx$ 620 nm). Coincidently, this matches well with the peak of the solar spectrum (Fig. 1.4).

A single catalyst powered by UV/Visible light can be efficient (Eg = 1.4 eV, $\eta$ = 30 %/Eg = 2.0 eV, $\eta$ = 10 %) under solar irradiation [10]. However, many catalysts still suffer from low activity due to incomplete charge separation. Photogenerated electron/hole pairs often recombine to form fluorescence or waste heat. In order to better evaluate the photocatalytic abilities of a catalyst in aqueous solutions, a sacrificial reagent can be added (Fig. 1.5). Sacrificial reducing agents (alcohol, $S^{2-}/S_8$, $Fe^{2+}/Fe^{3+}$, $I^-/IO_3^-$ | E < −0.82 V vs. NHE pH 7) are more easily oxidized than water and become oxidized by the catalyst instead of water. This removes a portion of the holes from the catalyst, leaving behind a higher concentration of electrons to boost hydrogen evolution. Sacrificial electron acceptors ($Ag^+/Ag^0$, $Ce^{4+}/Ce^{3+}$, $S_2O_8^{2-}/SO_4^{2-}$, $Fe^{3+}/Fe^{2+}$, $IO_3^-/I^-$ | E > + 0.41 V vs. NHE pH 7) on the other hand, are more easily reduced than water and remove electrons from the catalyst to liberate the photogenerated holes, and this promotes oxygen

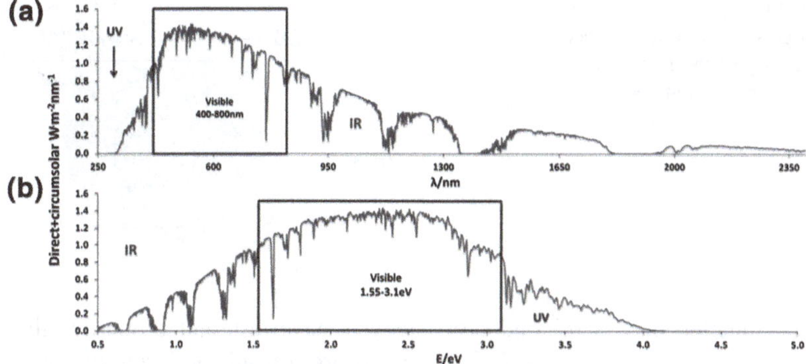


**Fig. 1.4** Average daily solar irradiance (AM 1.5) [16] containing Ultraviolet (*UV*), Visible, and Infrared (*IR*) light plotted against wavelength (**a**) and photon energy (**b**)

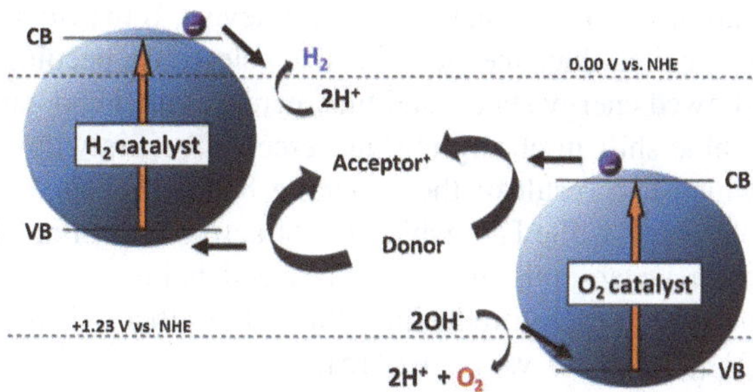


**Fig. 1.5** Sacrificial reagents for $H_2$ catalysts (Electron Donor) and $O_2$ catalysts (Electron Acceptor)

formation. Each of these chemicals are used to study only one of the electrolysis half reactions; however, positive results do not guarantee that a catalyst will be active for overall water splitting in the absence of sacrificial reagents [11].

Cocatalysts also enhance photocatalysis by lowering the activation energy/overpotential of the water redox reaction [11, 12]. Cocatalysts are generally deposited onto the semiconductor particle at a low weight percent (0.5–5 %) via photodeposition, drying aqueous salts, annealing particles at high temperatures, or by adding capped pre-synthesized particles. When considering that photocatalyst particles are miniature electrochemical cells, the attachment of small electrodes for water redox reactions can enhance charge transport to the electrolyte (Fig. 1.6). Metal cocatalysts with large work functions ($\phi > 4.70$ eV vs. vacuum) such as Pt, Pd, Ni, Ru, Rd form a rectifying Schottky diode with the catalyst and typically enhance electron trapping to improve $H_2$ evolution rates by serving as a cathode. Semiconducting metal oxide cocatalysts such as $IrO_2$ and $RuO_2$ form diodes with the catalyst and can enhance hole trapping to raise $O_2$ evolution rates by acting as the anode. In this way, inactive materials can be activated for photo-reduction or photo-oxidation of water in a two-component system or activated for both reactions in a three-component system [13].

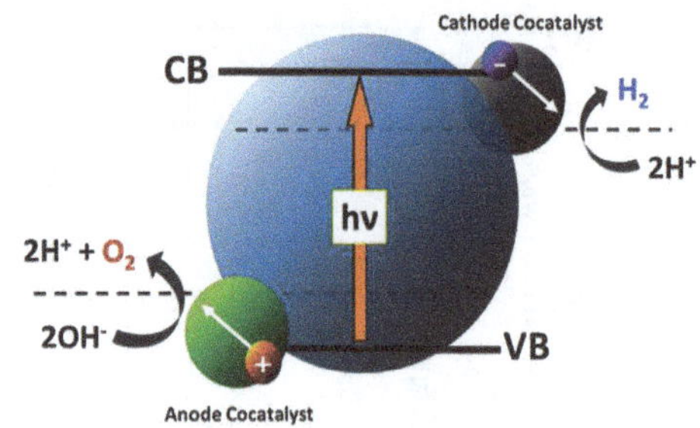


**Fig. 1.6** Cocatalyst effect on charge separation to enhance efficiency of the hydrogen evolution reaction and the oxygen evolution reaction

Nanometer sized semiconductor catalysts have different properties than their microscale bulk derivatives. These changes are a result of quantum confinement effects in addition to increased surface states. For example, the surface area of 1 μm spherical $Fe_2O_3$ particles (1.1 $m^2\ g^{-1}$) is increased by a factor of 200 for nanoparticles of 5.4 nm diameter. Increased catalytic surface leads to higher rates of activity, but can concurrently increase surface trap states that act as recombination sites or sub-band excitation levels. If the Bohr-exciton radius of the material is smaller than the particle size, electrons become quantum confined to fewer allowed energy states, and this increases the band gap energy. In the case of CdSe, a blue shift in absorption/fluorescence occurs with decreasing size below 4.9 nm which is a result of the widening band gap/band edges. This has been observed with bulk-CdSe [14] which is not active for colloidal photocatalytic $H_2$ evolution from water until its size is reduced below 3 nm. At this size, the CB energy becomes more reducing than the theoretical proton reduction potential ($E_{CB} < 0.00$ V vs. NHE) [15].

Newfound properties of nanomaterials offer improvements on microscale photocatalysts since their confined size contributes to tunable band gaps and improved charge carrier extraction. Since nanoscale photocatalyst studies are generally limited to thin-film electrodes in an electrochemical cell under applied bias, there are few studies of un-assisted colloidal suspensions (Type 1). Motivation for the research described here is fueled by the desire to begin to fill the void of knowledge about nanomaterials for energy conversion. This research explores new bottom-up avenues to well-defined nanoscale systems, systematic testing of photocatalytic activity, measurement of key parameters (atomic structures, morphologies, overpotentials, conduction band edges), to achieve a better understanding of limiting factors and arrive at improved systems for solar energy to fuel conversion.

## References

1. The-World-Factbook, *CIA* (2006)
2. U.S.-Energy-Information-Administration (2011)
3. J. Barber, Chem. Soc. Rev. **38**, 185–196 (2009)
4. N.S. Lewis, D.G. Nocera, Proc. Natl. Acad. Sci. U.S.A. **104**, 20142 (2007)

5. W.U. Huynh, J.J. Dittmer, A.P. Alivisatos, Science **295**, 2425–2427 (2002)
6. U. Bach, D. Lupo, P. Comte, J.E. Moser, F. Weissortel, J. Salbeck, H. Spreitzer, M. Gratzel, Nature **395**, 583–585 (1998)
7. M. Gratzel, Nature **414**, 338–344 (2001)
8. G.N. Baum, J. Perez, K.N. Baum, DOE **1**, 1–127 (2009)
9. A. Fujishima, K. Honda, Nature **238**, 37 (1972)
10. F.E. Osterloh, B.A. Parkinson, MRS Bull. **36**, 17–22 (2011)
11. K. Maeda, J. Photochem. Photobiol. C **12**, 237–268 (2011)
12. F.E. Osterloh, Chem. Mater. **20**, 35–54 (2008)
13. T.K. Townsend, E.M. Sabio, N.D. Browning, F.E. Osterloh, ChemSusChem **4**, 185–190 (2011)
14. F.A. Frame, E.C. Carroll, D.S. Larsen, M. Sarahan, N.D. Browning, F.E. Osterloh, Chem. Commun. **19**, 2206–2208 (2008)
15. M.A. Holmes, T.K. Townsend, F.E. Osterloh, Chem. Commun. **48**, 371–373 (2012)
16. ASTM G173-03 Reference Spectra Derived from SMARTS v. 2.9.2. Data provided by The National Renewable Energy Laboratory, Golden Colorado (2008), http://www.nrel.gov/

# Chapter 2
# The Hydrogen Evolution Reaction: Water Reduction Photocatalysis—Improved Niobate Nanoscroll Photocatalysts for Partial Water Splitting

## 2.1 Introduction

Development of a photocatalyst for harvesting the practically endless sources of fuel from sunlight and water may be one of the most important challenges of the century. Semiconductor photocatalysts are expected to be the most promising solution to the global energy dilemma [1]. However, due to low efficiencies ($\eta < 10\ \%$) of solar to hydrogen conversion and limited light absorption, no economically competitive water splitting photocatalyst has yet been realized. Compared to the traditionally large catalysts (>1 μm), nanomaterials offer potential advantages due to their higher catalytic surface areas and shorter semiconductor-liquid electron pathways for charge and exciton transport. For example, Bulk $Fe_2O_3$, which is known to suffer from high recombination rates and short diffusion lengths was made active as a photoanode nanomaterial [2]. In addition to enhanced surface areas, nanoscale semiconductors also exhibit quantum confinement effects which can be used to adjust energy levels and light absorption. This phenomenon allowed for non-active bulk-CdSe to be activated on the nanoscale because of the raising of the conduction band above the electrochemical reduction potential for water [3, 4]. One problem with nanostructures that needs to be addressed is that the space charge regions, if present at all, are not strong enough to enable efficient electron hole separation. The attachment of cocatalysts could potentially solve this issue since they can act as selective electron hole acceptors. However, data on the energetic structures of these nanomaterials and composites are still scarce, so optimization of such contacts presents a challenge. In addition, there are also synthetic obstacles for the preparation of well-defined multi-component nanostructures. As part of our ongoing search for a nanoscale water splitting photocatalyst [3–12], we describe here scalable syntheses of catalysts that contain up to three separate nanoparticle components for light absorption, water reduction and oxidation. For this investigation, we used a nanoscale derivative of

This chapter appeared as: "T.-K. Townsend, et al., *Improved Niobate Nanoscroll Photocatalysts for Partial Water Splitting*, Chem. Sus. Chem. **4**(2), 185–190 (2011)."

T. K. Townsend, *Inorganic Metal Oxide Nanocrystal Photocatalysts for Solar Fuel Generation from Water*, Springer Theses, DOI: 10.1007/978-3-319-05242-7_2,

$K_4Nb_6O_{17}$ which is known to exfoliate into an asymmetrical scroll structure as the light absorbing material [13–16]. Charge separation on this material is thought to also be asymmetric due to the nature of the unique crystal orientations of the scroll framework. In order to introduce electronically separated sites for water reduction and oxidation, Pt and IrOx (x = 1.5–2) nanoparticle cocatalysts are attached to the nanoscrolls via photochemical deposition, respectively. The morphology, optical, electrochemical, photocatalytic and energetic properties of these nanomaterials are also investigated.

## 2.2 Results and Discussion

The formation of niobate nanoscrolls and the incorporation of the Pt and IrOx cocatalysts are shown in Scheme 2.1 (Eqs. 2.1–2.4). After the solid state synthesis of bulk $K_4Nb_6O_{17}$ crystals, dilute acid and tetrabutylammonium hydroxide addition leads to the exfoliation of the crystal into individual nanosheets. These naturally turn into scrolls **[1]** because of the inherent strain of the asymmetrical structure [16]. When **[1]** are irradiated in the presence of potassium hexachloroiridate (III) and nitrate as electron acceptor, one obtains the scrolls modified with IrOx cocatalyst **[2]** (Scheme 2.1, Eq. 2.2). Scrolls decorated with platinum nanoparticles (Scheme 2.1, Eq. 2.3) can be produced via photochemical reduction of hexachloroplatinic acid and methanol as an electron donor **[3]**. By combining these two cocatalyst depositions (Scheme 2.1, Eq. 2.4), three-component nanocatalysts were produced **[4]** by either sequential deposition of first Pt and then IrOx, or by co-deposition in a one-pot deposition. Products of these aqueous processes were precipitated and cleaned as described in the experimental section.

$$K_4Nb_6O_{17(s)} + 4HNO_{3(aq)} + 4(Bu_4N)OH_{(aq)} \xrightarrow{5\ days} \mathbf{(Bu_4N)_4Nb_6O_{17(s)}}[1] + 4KNO_3 + 4H_2O \tag{2.1}$$

$$(Bu_4N)_4Nb_6O_{17(s)} + K_3IrCl_6 + {}^1\!/_2KNO_3 + 3OH^- \xrightarrow{hv} [\mathbf{1wt.\%}\ \boldsymbol{IrO_2}](Bu_4N)_4Nb_6O_{17}[2] + {}^1\!/_2KNO_2 + {}^3\!/_2H_2O + 3K^+ + 6Cl^- \tag{2.2}$$

$$(Bu_4N)_4Nb_6O_{17(s)} + H_2PtCl_6.6H_2O + 6OH^- \xrightarrow{hv} [\mathbf{1wt.\%}\ \boldsymbol{Pt}](\boldsymbol{Bu_4N})_4\boldsymbol{Nb_6O_{17}}\ [3] + O_2 + 6Cl^- + 10H_2O \tag{2.3}$$

$$[4], \boldsymbol{wt\%} = \frac{\boldsymbol{mass\ Pt\ or\ IrO_2}}{\boldsymbol{mass\ Bu_4N\ _4Nb_6O_{17}}} \boldsymbol{\times 100}$$

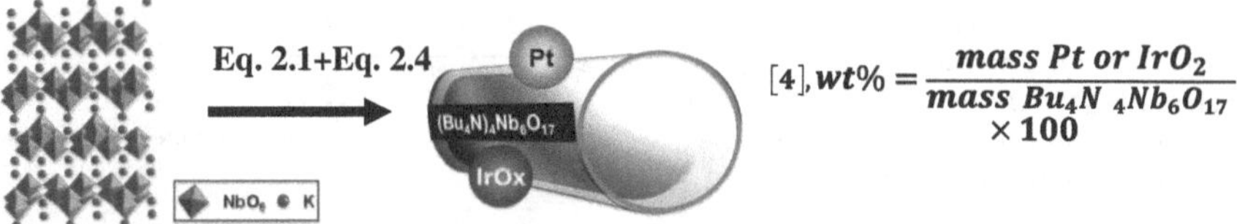

**Scheme 2.1** Synthesis of nanocrystal photocatalysts (Eqs. 2.1–2.4). The structure of $K_4Nb_6O_{17}$ (including exfoliation *lines*) and the schematic structure of the nanoscrolls with co-catalysts are also shown where wt.% cocatalyst loading is defined

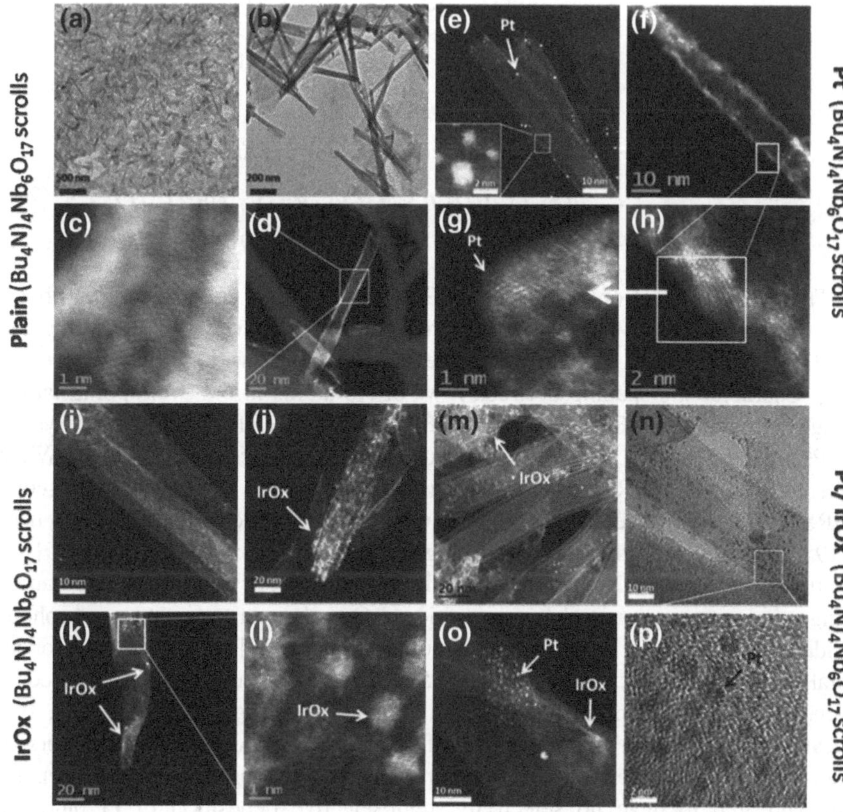

**Fig. 2.1** TEM (**a**, **b**), High Angle Annular Dark Field (HAADFSTEM) (**c–m**, **o**), and HRTEM (**n**, **p**) of $(Bu_4N)_4$ $Nb_6O_{17}$ nanoscroll (**a–d**), Pt nanoscroll (**e–h**), IrOx nanoscroll (**i–l**), and Pt/IrOx nanoscroll (**m–p**)

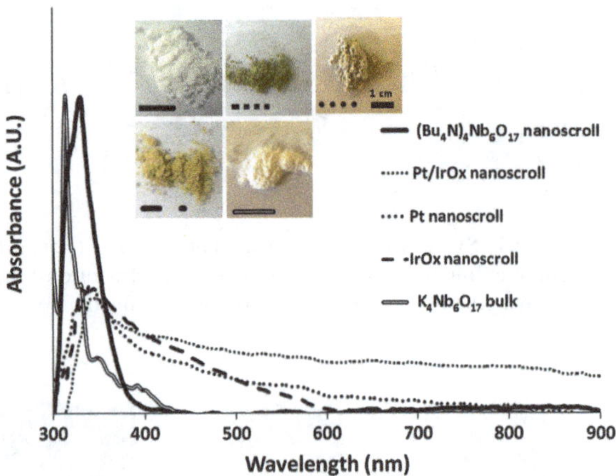


**Fig. 2.2** Diffuse reflectance UV/Vis spectra of four catalysts and the parent material ($K_4Nb_6O_{17}$) with photographs showing coloration due to cocatalyst deposition

$$\begin{aligned} &(Bu_4N)_4Nb_6O_{17(s)} + K_3IrCl_6 + H_2PtCl_6.6H_2O + 1/2KNO_3 + 9OH^- \\ &\quad \overset{hv}{\rightarrow} [\boldsymbol{0.5wt.\%IrO_2|0.5wt.\%Pt}](\boldsymbol{Bu_4N})_4\boldsymbol{Nb_6O_{17}[14]} \\ &\quad + 1/2KNO_2 + 3K^+ + 11.5H_2O + O_2 + 12Cl^- \end{aligned} \tag{2.4}$$

Transmission electron microscopy (TEM) was used to confirm the morphologies of the products **1–4**. The niobate scroll platform [**1**] are 1 ± 0.5 μm long and 10 ± 5 nm wide (Fig. 2.1a, b), in agreement with observations by Saupe et al. [16]. Characteristic spacing between rows of Nb atoms in the overlapping lattice fringes are evident in Fig. 2.1c, d [17, 18]. The surfaces of [**3**] show nanocrystals of $Pt^0$ metal (<2 nm) discretely deposited on the surface (Fig. 2.1e–h). When imaged with High Resolution Scanning Transmission Electron Microscopy (HRSTEM), the lattice fringes of the crystalline Pt particles were measured as 0.195 nm and match the known spacing of platinum for the [100] axis [19]. In contrast, the 1 wt.% $IrO_x$ nanoscrolls in Fig. 2.1i–l showed no visible lattice fringes indicating an amorphous deposition. $IrO_x$ is known to form an amorphous oxide/hydroxide phase when grown in solution phase [20]. The three-component catalysts containing 0.5 wt.% Pt and 0.5 wt.% IrOx share features displayed in both of the two-component structures. Pt and IrOx (Fig. 2.1m–p). These two cocatalysts could also be directly identified and mapped with Energy Dispersive Spectroscopy (EDS, see Fig. 2.7 in Supporting Information). While IrOx deposition appeared to have no selectivity, regioselective deposition of Pt onto the asymmetric nanoscroll was observed with Pt growing predominately on the edges of the rolled crystal sheets.

The optical absorption edge of $K_4Nb_6O_{17}$ (348 nm) corresponds to the 3.55 eV band gap of the parent material is shown in the UV-Vis spectra (Fig. 2.2). The nanoscale version of this material [**1**] has a red-shifted absorption onset (to 375 nm) corresponding to a reduced band gap of 3.3 eV which we have previously attributed to Nb–O bond weakening as a result of strain induced from bending of

**Table 2.1** Catalyst turnover numbers calculated as moles of formed $H_2/O_2$ per mole of $H_4Nb_6O_{17}$ catalyst (MW = 834 g/mol, 120 μ mol for 100 mg) after a 5 h irradiation

| | Catalyst | Solution | | |
|---|---|---|---|---|
| | $H_4Nb_6O_{17}$ nanoscroll | Pure water | 20 % (vol) methanol | 25 mM $AgNo_3$ |
| (1) | Unmodified | 2.3 | 4.9 | 0.1 |
| (2) | IrOx | 8.3 | 6.24 | 0.1 |
| (3) | Pt | 8.3 | 18.5 | 0.0 |
| (4) | Pt/IrOx | 4.6 | 14.4 | 0.1 |

the crystal plane [9]. After cocatalyst deposition, the materials become colored (see photographs in Fig. 2.2) due to the visible absorption of Pt and IrOx. The Pt coated scrolls appear brown due to broad visible absorption similar to that of Pt colloids [21]. The $IrO_x$ coated scrolls are also brown, not blue as observed for colloidal Ir(IV)$O_2$ [22]. The brown color indicates the presence of $Ir^{3+}$ oxidation state which may not be as active as $Ir^{4+}$; however, Ir(III)oxide is also known to be an active water oxidation catalyst [23, 24]. The three component samples share optical characteristics with both two-component catalysts.

Photo-irradiation experiments with the nanoscroll catalysts were conducted over 5 h periods in aqueous methanol (20 % vol), 25 mM aqueous silver nitrate solution, and in pure water, using 100 mg (~55.6 μ mol) of catalyst in each case. From methanol solution, all catalysts evolve $H_2$ at a constant rate. [Pt] $(Bu_4N)_4$ $Nb_6O_{17}$ **[3]** is most active (~500 μ mol/h), followed by [IrOx, Pt] $(Bu_4N)_4$ $Nb_6O_{17}$ (**4,** ~250 μmol/h), $(Bu_4N)_4$ $Nb_6O_{17}$ **[1]**, and [IrOx] $(Bu_4N)_4$ $Nb_6O_{17}$ **[2]** the latter of which evolve $H_2$ at nearly the same rate of ~100 μmol/h. Turnover numbers range between 10.6 (for **1**) to 40 (for **3**), establishing a catalytic process (Table 2.1).

Each catalyst **[1–4]** evolve $H_2$ from pure water in the presence of light, but the rates are only about half of that in aqueous methanol. As expected, **[2]** shows the lowest activity in methanol and the relative activities of the four catalysts remain unchanged whether in aqueous methanol or in water. Importantly, we find that in water, $H_2$ evolution cannot be sustained and in the case of the most active **[3]** drops to nearly zero after 24 h of continuous irradiation (Fig. 2.3d). We have previously attributed this to the formation of surface peroxides which terminate the oxygen formation reaction and inhibit the active sites for water reduction as seen with the related phase $HCa_2Nb_3O_{10}$ [8]. The formation of hydrogen and Nb (5 +)-bound $\mu^2$-peroxide is favored over the more demanding water oxidation steps. To investigate this further, titration of pre-irradiated platinated scrolls **[3]** was performed with o-Tolidine indicator (details in supporting information) revealing that peroxides are also present in 1:1 stoichiometry with regard to the amount of gaseous $H_2$ released (Fig. 2.3d) confirming the surface passivation of the catalysts with peroxides. After replacement of residual tetrabutylammonium (TBA) surfactant with protons via repeated precipitation/washing steps, the $H_4Nb_6O_{17}$ scrolls (100 mg) evolve up to 680 μmol of $H_2$ over a 24 h period, i.e. $H_2$:$H_4Nb_6O_{17}$ = 5.7. This number

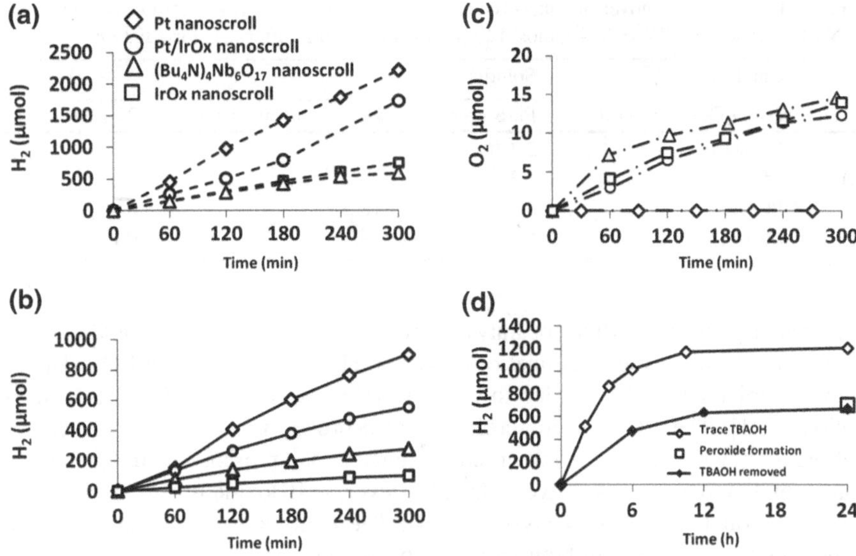


**Fig. 2.3** $H_2$ and $O_2$ evolution under UV-light irradiation with a 300 W Xe arc lamp from 100 mg of catalysts suspended in 50 mL of 20 % aqueous methanol (**a**– –), pure water (**b**—), and 25 mM aqueous $AgNO_3$ (**c**–·–). (D). Hydrogen evolution during long term (24 h) irradiation of 100 mg 1 %Pt nanoscrolls (3) in water. Also shown in the plot is $H_2$ evolution data for 3 after removal of residual tetrabutyl ammonium hydroxide (TBAOH), and the amount of peroxide detected coulometrically via reaction with o-tolidine (for details see Fig. 2.8 in Supporting Information)

agrees very well with our experimental data assuming that each of the six niobium ions in the scrolls can accommodate one peroxide ligand, the scrolls can release at most six mole equivalents of hydrogen. In the presence of TBA, the amount of $H_2$ from irradiation in water is higher (see Fig. 2.4d and first column in Table 2.1). We attribute this to side reactions of the peroxide with the tetrabutylammonium ion, as previously described for TBA $Ca_2Nb_3O_{10}$ nanosheets [8].

In contrast to $H_2$ evolution, water oxidation to promote $O_2$ evolution is a multistep process involving high energy intermediates. As a result, peroxide intermediate products are formed without further reaction. This leads to difficulty of the photocatalysts to release $O_2$ from water and even from sacrificial solution of aqueous $AgNO_3$ (Fig. 2.3). The $O_2$ evolution rate declines over time, in part due to peroxide poisoning of the catalyst, and also as a result of Ag particle deposition onto the scrolls. Only non-catalytic amounts of $O_2$ are formed, or none as in the case of [Pt] $(Bu_4N)_4$ $Nb_6O_{17}$.

Cyclic voltammetry was used to gain further insight into the reasons for varying catalytic activities of these nanocomposites (Fig. 2.4). Electrochemically, water oxidation is observed for all materials significantly above the theoretical potential (+820 mV vs. NHE at pH 7). The first oxidation wave at +1.10 belongs to the two-electron oxidation of water to peroxide as shown in Fig. 2.5, with a corresponding

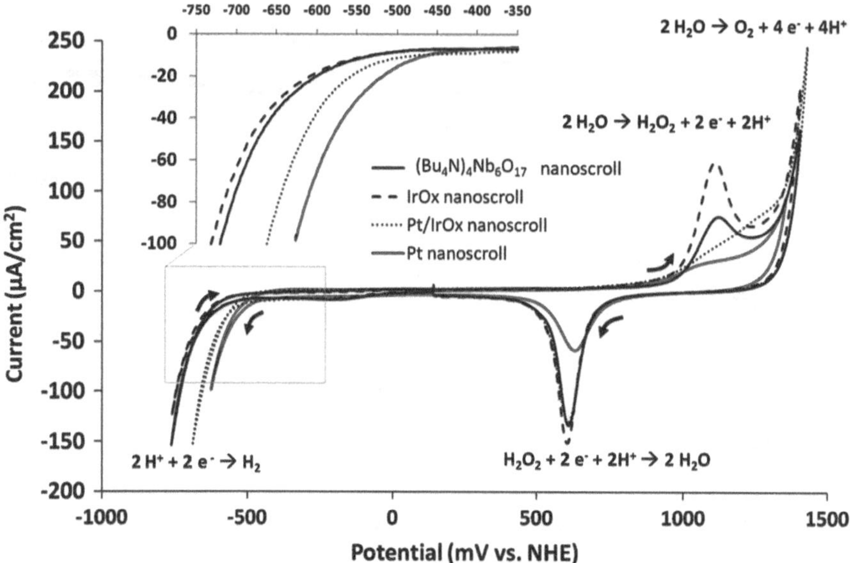


**Fig. 2.4** Cyclic voltammograms of films of the four catalysts on a Au electrode (1 $cm^2$) in 0.25 M $NaH_2PO_4/Na_2HPO_4$ buffer at pH 7 and scan rate of 10 mV/s. *Arrows* show scan direction. All potentials are given versus NHE. Additional scans on bare Pt and Au electrodes (Fig. 2.9 in supporting information) show that background currents from the Au electrode can be neglected

**Fig. 2.5** Correlation between measured reduction overpotentials and photocatalytic $H_2$ evolution rate from 20 % (vol) aqueous methanol (– –) and pure water (—). Error bars correspond to 10 % experimental error

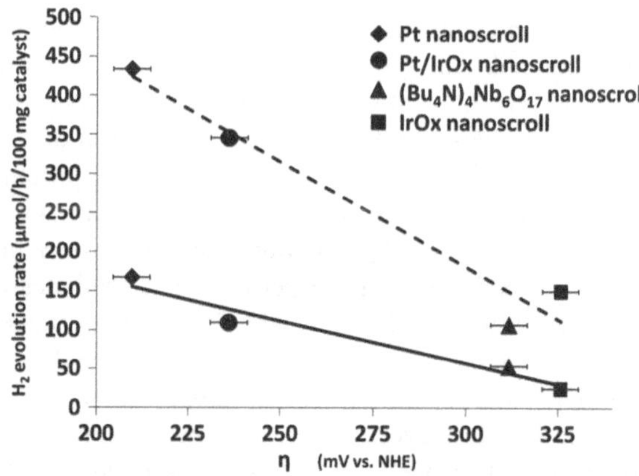


reduction process on the reverse scan occurring at +600 mV. The shapes and potentials of these redox features are very similar to gold electrodes, after correcting for the difference in pH [25, 26]. Indeed, if the nanoscroll film is kept at 1.2 V NHE bias for 1 h, peroxides can be directly detected on the surface using a peroxidase-coated test strip (Fig. 2.10). Peroxide formation is favored energetically because of the strong binding affinity to Nb (5+) and kinetically, because it

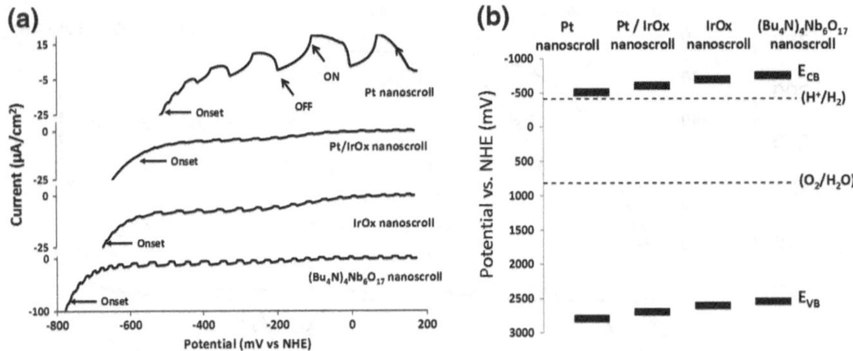


**Fig. 2.6** Reduction scans (10 mV/s) of catalysts deposited and dried as a thin film onto a gold 1 $cm^2$ electrode in 25 mM $NaH_2PO_4$ pH 7 buffer with chopped light (**a**) and resulting energy band diagrams of each catalyst (**b**)

involves fewer redox steps and fewer reactive intermediates in comparison to $O_2$ formation.

Cathodic scans reveal significant differences of 0.21–0.325 V between the actual and theoretical reduction potential for water (−410 mV vs. NHE at pH 7). The scrolls containing 1 % Pt cocatalysts **[3]** showed the lowest reduction overpotential (0.21 V), as would be expected from the high catalytic conversions. For the other catalysts, cathodic overpotentials increase in the order Pt/IrOx **[4]** > nanoscroll **[1]** > IrOx **[2]**. This parallels the order of $H_2$ evolution activity for both water and aqueous methanol reactions. When plotted against the electrochemical overpotentials, the rates of $H_2$ evolution can be seen to fit a straight line in both cases, although the slope is greater in the presence of methanol (Fig. 2.5). This is expected because methanol acts as an electron donor since methanol oxidation is more favorable than water oxidation, so the proton reduction rate becomes the limiting factor. In pure water, the water oxidation reaction becomes rate limiting, based on the lower $H_2$ evolution rates from water in Fig. 2.2.

Water oxidation to form peroxide intermediates is observed as a peak for **[1]** and **[2]** and as a shoulder for the Pt containing materials, which could indicate a fast subsequent oxidation reaction (i.e. Pt—catalyzed peroxide disproportionation into $H_2O$ and $O_2$. Based on the observation of residual peroxide reduction at +600 mV on the reverse scan, the consumption of peroxides is also incomplete for these catalysts as well, albeit lower amplitude for **[3]** with the most Pt loading. On the other hand, IrOx loaded scrolls are nearly identical to the bare scrolls which indicates that water oxidation proceeds mainly on the surface of the niobate rather than the IrOx sites as proposed.

Gas bubbles were observed evolving from the surface of the electrode as the voltage scan reached >1400 mV, corresponding to the formation of $O_2$. This is roughly the same for each catalyst, indicating that IrOx assumes a spectator role, and that the water oxidation reaction occurs on the surface of the niobate. This could be due to a few factors including the low mass fraction of the IrOx in **[2]** and that there does not appear to be an efficient pathway for hole transport from the Nb–O sites to the IrOx particles. Also the amorphous IrOx may be less active for water oxidation in the +3 state versus the blue $IrO_2$ [23, 24]. Given that the water oxidation overpotential ($\sim$600 mV) is two–three times larger than the proton reduction overpotential (210–325 mV), it is clearly the oxygen evolution reaction that is the limiting factor in the photocatalytic water splitting activities of these materials. Since the Ag ion reduction potential of +0.8 V [27] is simply not sufficient to promote $O_2$ evolution from the scrolls, this also explains the low efficiency of $O_2$ evolution in the presence of silver nitrate.

Chopped light voltammetry experiments were also performed in the presence of methanol as an electron donor (Fig. 2.6a) in order to determine the onset potentials and the magnitude of the photocurrents. Each catalyst responds to illumination with a positive photocurrent, and the onset of this can be used to estimate the conduction band edge in highly n-doped semiconductors [28–30]. Onset potentials are observed from $-0.50$ to $-0.80$ V at pH $= 7$, and the sizes of the photocurrents differ among catalysts with the strongest currents observed with [Pt] $(Bu_4N)_4Nb_6O_{17}$ [**3**]. Together with the optical band gaps, this data was used to construct the energy diagrams in Fig. 2.6b. Based on this information, we can see that the conduction band energies of each catalyst are above the theoretical proton reduction potential in neutral solution (–410 mV vs. NHE) and therefore active for $H_2$ production. The addition of cocatalysts was found to depress the conduction band edge by 58 mV (IrOx), 148 mV (Pt/IrOx), and 242 mV (Pt) in positive direction from the value for the unmodified scrolls ($-748$ mV). We observed similar results with Pt-modified $H_2Ti_4O_9$ nanocrystals [12] and this can be interpreted as a result of electron transfer from the niobate to the cocatalysts in the direction of lower Fermi energy.

Due to the enormous interfacial area of the nanoscrolls and the cocatalyst particles, this phenomenon is more pronounced than would be expected for bulk materials. One might expect that the associated polarization should lessen the thermodynamic driving force for water reduction, thereby reducing the $H_2$ evolution rate. However, the opposite is observed here, because the photocatalytic $H_2$ activity in methanol solution is not controlled by the energetics of the catalysts but by the kinetics of proton reduction (electrochemical overpotentials). For water oxidation the situation is similar. Even though the valence band edges of all catalysts are well below the water oxidation potential, oxygen formation is inhibited kinetically due to the large overpotential for its formation and to the greater ease of peroxide formation on the niobate surface.

## 2.3 Conclusion

We have synthesized two- and three-component photocatalysts by a combination of top-down (exfoliation from bulk) and bottom-up approaches (photochemical deposition). The resulting Pt- and/or IrOx coated niobate nanoscrolls evolve catalytic amounts of $H_2$ from aqueous methanol solutions but only limited amounts from pure water due to the formation of peroxide intermediates. This is also seen from the non-stoichiometric $O_2$ evolution from aqueous $AgNO_3$. In methanol, the $H_2$ evolution rates are inversely correlated with $H^+$ overpotentials, which shows that the process is limited by the rate of proton reduction on Pt. In water and aqueous silver nitrate, oxygen is inhibited due to the large electrochemical overpotential (2–3 times larger than the $H_2$ overpotential) that acts as a kinetic barrier for $O_2$ formation. Instead, peroxide intermediates are favored due to the strong binding affinity of surface peroxides to Nb (5+). The addition of Pt and IrOx cocatalysts do not prevent the formation of niobate-bound peroxides and leads to diminishing hydrogen evolution from water. We also find that cocatalyst attachment causes a depression of the conduction band energy based on the photo-onset potentials. However, this does not affect the activity of the catalysts in the presence of other more important factors.

## 2.4 Experimental

Reagents: $K_2CO_3$, $Nb_2O_5$, and TBA (OH) (40 wt.% in $H_2O$) were purchased from Acros Organics and used Scheme 2.1. $H_2PtCl_{-6}$ x$6H_2O$, $K_3IrCl_6$, $H_2SO_4$ and $HNO_3$ (70 %) were purchased from Sigma-Aldrich. $KNO_3$ and $AgNO_3$ were obtained from Fisher Scientific, Pittsburgh, PA. Reagents were of reagent quality and were used as received. Water was purified to >18 M Ω cm resistivity using a Nanopure system. $K_4Nb_6O_{17}$ was prepared according to a literature method [13].

**$(Bu_4N)_4[Nb_6O_{17}]$ scrolls [1].** The scrolls were prepared as described previously [16]. Initially, an acid treatment of 2.0 g (1.93 mmol) bulk $K_4Nb_6O_{17}{\cdot}3H_2O$ powder with 250 mL of 2 M $HNO_3$ was carried out for 5 days. The resulting white product was centrifuged and washed four times with 50 mL of pure water. Exfoliation of the wet acid exchange product involved adding TBA hydroxide (40 wt.% in water, Acros) in a 20:1 molar ratio (26.6 mL, 40.5 mmol) and stirred for 5 days. This white solution was centrifuged and washed twice with pure water. During the second wash, the precipitate was discarded and the supernatant (containing a homogeneous colloid suspension of nanoscrolls) was isolated for the following study. The concentration of the colloid was measured gravimetrically to be 10 mg/mL which corresponds to an average total of 1.0 g $(Bu_4N)_4$ $[Nb_6O_{17}]$ (28 % yield, M = 1798 g/mol). Transmission electron microscopy (TEM) reveals thin sheets that have rolled into a rod-like structure with multiple layers.

**[Pt, IrOx, and Pt/IrOx] $(Bu_4N)_4[Nb_6O_{17}]$ scrolls [2, 3, 4].** Platinum nanoparticles were deposited onto the surface of the niobate scrolls by adding 2.13 mg $H_2PtCl_6$ (85 wt.% in water, Acros) to a dispersion of 100 mg of niobate scrolls in 50 mL of water. The solution was purged with Ar gas and irradiated under UV-light for 1 h with a Xe-arc lamp. The colored product **3** (Fig. 2.2) was precipitated with 0.1 M $H_2SO_4$, washed twice with 50 mL of pure water and then re-dispersed in 0.5 wt.% tetrabutyl-ammonium hydroxide in water. Bright field TEM images reveal the Pt nanoparticles as dark spots (2–3 nm in diameter) containing the characteristic lattice fringes of platinum [19]. Iridium oxide ($IrO_x$) modified nanoscrolls [**2**] were prepared analogously, using 2.33 mg of $K_3IrCl_6$ (99.9 % purity, Acros) and 5 mM (25.25 mg) $KNO_3$ for 100 mg of niobate scrolls dispersed in 50 mL of water. The nitrate is known to facilitate deposition of IrOx [31]. Scanning TEM (STEM) images show additional broader regions of high contrast corresponding to $IrO_x$ deposition. Nanoscrolls containing both 0.5 % Pt and 0.5 % IrOx [**4**] were synthesized in two ways. Pt and IrOx were photodeposited sequentially, with a water washing step (50 mL) after the first deposition. Alternatively, Pt and IrOx were deposited simultaneously using a mixture of 1.17 mg $K_3IrCl_6$ and 1.07 mg $H_2PtCl_6$ in 50 mL of 5 mM $KNO_3$ before washing four times with 50 mL of water. The two products were found to be identical in terms of their morphology and their optical and catalytic properties.

The rate of photochemical hydrogen evolution from each catalyst was determined by irradiating 100 mg of each catalyst in 50 mL of pure water in a quartz round bottom flask with a 300 W Xe-arc lamp (209 mW/cm$^2$ at flask, $\lambda = 220$–380 nm, measured with GaN photodetector). The air-tight irradiation system connects a vacuum pump and a gas chromatograph (Varian 3800) with the sample flask to quantify the amount of gas evolved, using area counts of the peaks and the identity of the gas from the calibrated carrier times. Prior to irradiation, the flask was evacuated down to 5 torr and purged with argon gas. This cycle was repeated until the chromatogram of the atmosphere above the solution indicated that the sample did not contain hydrogen, oxygen, or nitrogen. Electrochemical measurements were conducted by drying a drop of the aqueous dispersions of the catalyst under nitrogen gas onto a gold electrode (1 cm$^2$) and submerging the dried product into an air-tight quartz flask filled with 50 mL of 0.25 M $Na_2HPO_4$/$NaH_2PO_4$ buffer solution at pH 7, containing 20 % (volume) of methanol as sacrificial donor. The electrochemical cell consisted of a platinum counter electrode, the working electrode (catalyst), and a saturated calomel reference electrode connected to the cell with a KCl salt bridge. Nitrogen gas was bubbled through the electrolyte solution for 10 min prior to each experiment and remained streaming above the solution during the experiment. Photocurrents were measured by irradiation delivered via a fiber optics cable from a 300 W Xe arc lamp equipped with an IR filter. Chopped light with an irradiance of 15 mW/cm$^2$ at 200–380 nm illuminated the working electrode. The system was calibrated using the redox potential of $K_4[Fe(CN)_6]$ at +0.358 V (NHE). UV/Vis diffuse reflectance spectra were taken on white Teflon tape using an Ocean Optics HR 2000 CG-UVNIR spectrometer and a DH 2000 light source. Centrifugation was achieved with a IEC Centra CL2 centrifuge. Transmission electron images were taken with a Philips CM-12 instrument at 120 kV acceleration potential. Bright field high resolution transmission electron microscopy (BF-HRTEM) images were taken using a JEOL 2500SE 200 kV TEM. Z-contrast high angle annular dark field scanning TEM (HAADF-STEM) images were taken using a JEOL 2100F STEM with 200 kV field emission gun and a spherical aberration corrector.

## 2.5 Supporting Information

Figures 2.7, 2.8, 2.9, 2.10 and 2.11.

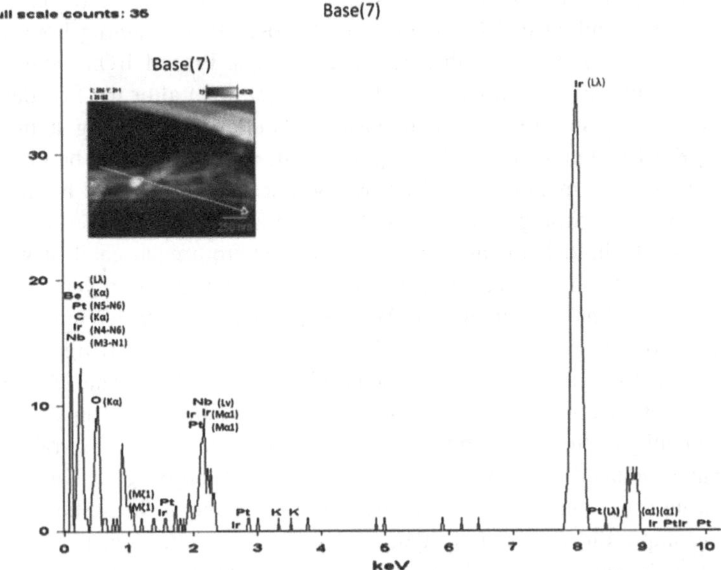


**Fig. 2.7** Energy Dispersive X-ray (EDX) line scan spectra of 1:1 Pt/$IrO_x$-loaded nanoscrolls. The spectra show multiple lines of Pt and Ir indicating their presence directly on the niobate. EDX was performed using a JEOL JEM 2500SE (S)TEM equipped with an EDX system

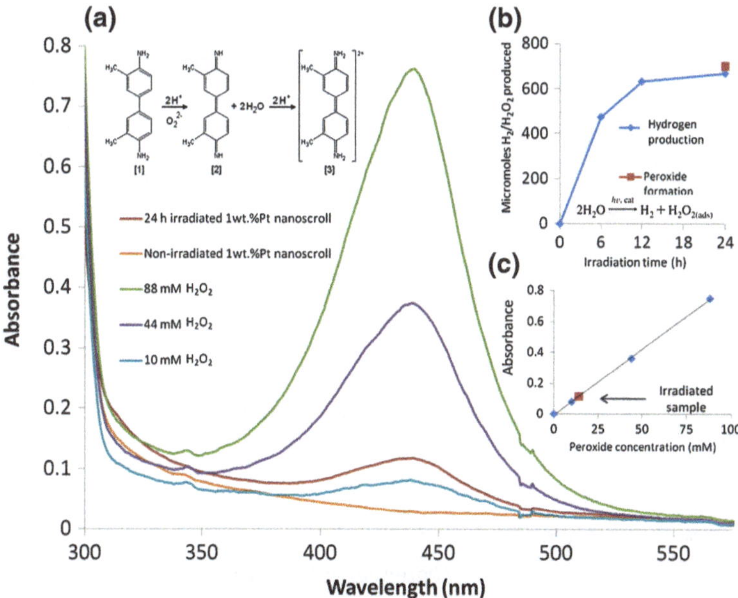


**Fig. 2.8** Peroxide determination with o-tolidine. UV-Vis spectra showing absorbance peak at 438 nm of protonated o-tolidine with varying concentrations of hydrogen peroxide plotted with an experimental 24 h water-irradiated nanoscroll sample mixed with o-tolidine. **a** the conversion of o-tolidine from the initial *white* product [1] which reacts with peroxide to form the *blue* product [2] and forms the protonated *yellow* species in acid [3]. **b** time-resolved hydrogen production from 100 mg of 1 wt.% Pt nanoscroll and the corresponding peroxide formation from an aliquot of the catalyst after irradiation. **c** the calibration curve for o-tolidine with hydrogen peroxide standards plotted with the experimental sample. Experiment follows procedure described in [8]

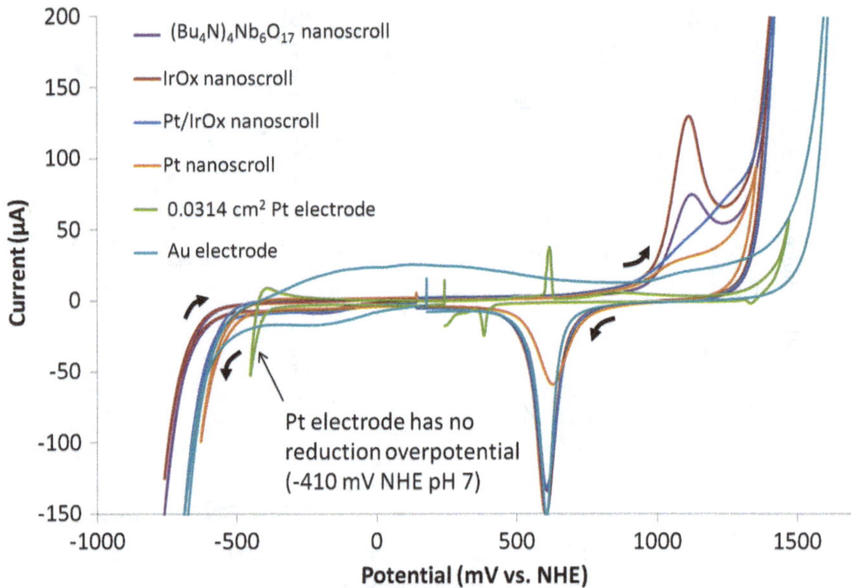


**Fig. 2.9** Electrochemical behavior of gold, platinum, and catalyst-coated electrodes. Cyclic voltammetry of four nanoscroll samples deposited as a thin film on a 1 $cm^2$ gold electrode in 50 mL of pH 7 $NaH_2PO_4/Na_2HPO_4$ buffer solution with scan rate of 10 mV/s including scans from bare Au electrode and Pt electrode for reference. *Arrows* show scan direction. Both Pt and Au electrodes oxidized water at potentials beyond +1.45 V. This rules out interference with the redox potentials observed for the catalysts, even in regions of partial electrode coverage with catalyst films. As expected, the bare Pt electrode facilitated proton reduction at −410 mV NHE @ pH 7, close to the proton reduction normal potential

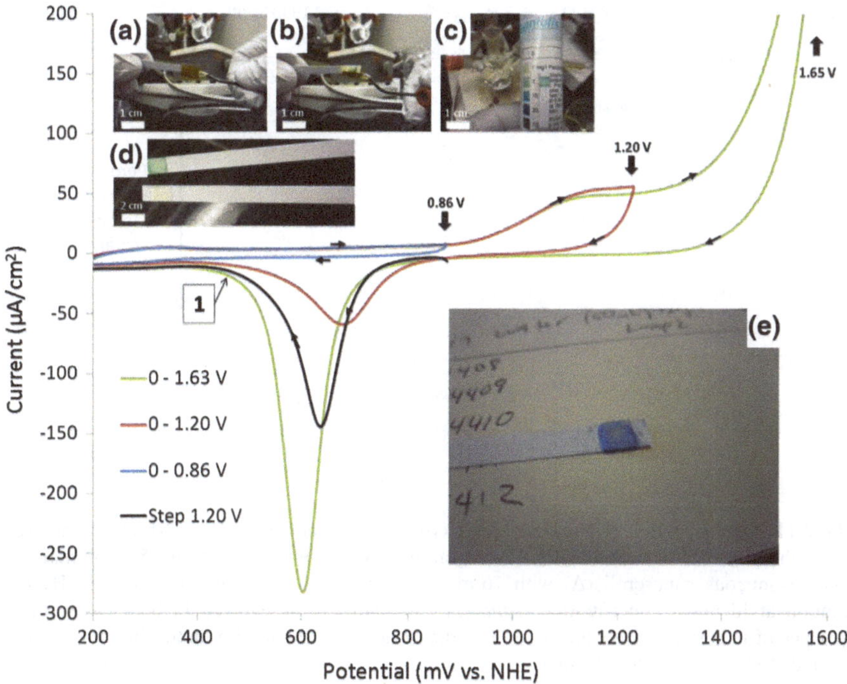


**Fig. 2.10** Peroxide detection on gold and niobate-coated electrodes. Cyclic voltammogram of bare gold 1 $cm^2$ electrode in 0.25 M $Na_2HPO_4/NaH_2PO_4$ pH 7 buffer with 10 mV/s scan rate. *Arrows* denote scan direction. *Blue*, *red* and *green curves* are components of a three-cycle hysteresis from 0.0 → 0.86 → 0.0 → 1.20 → 0.0 → 1.65 → 0.0 V. The *black curve* from 0.86 → 0.0 V was recorded after a chronoamperometry potential step voltage at 1.2 V for 100 min. Insets **a–d** shows qualitative peroxide determination via rubbing a *Quantofix peroxide 100* test strip on the gold electrode surface after 1.2 V chronoamperometry, where **a–b** show colored peroxide reaction, **c** approximate peroxide concentration ~3 mg/L, and **d** compares test after 1.2 V potential step (*blue*) and after reduction of peroxides (*white*, no reaction) at 500 mV [point 1]. Similar results were obtained for a film of $(Bu_4N)_4$ $Nb_6O_{17}$ on gold after an electrochemical bias at 1.2 V (NHE) for 1 h (insert **e**)

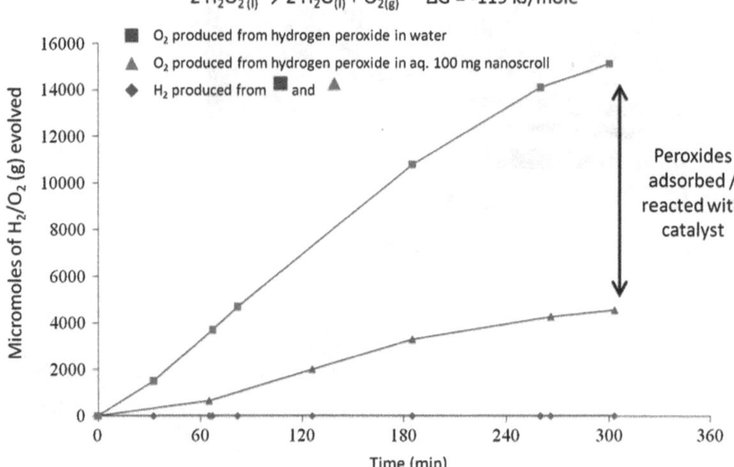


**Fig. 2.11** Demonstration of photocatalytic oxygen formation from $(TBA)_4$ $Nb_6O_{17}$ Nanoscrolls (1) in the presence of added $H_2O_2$. Hydrogen and oxygen evolution during 5 h irradiation of 100 mg aqueous nanoscroll (Δ) with 16 mmol of $H_2O_{2\ (l)}$ (*triangles*) and for pure $H_2O_{2\ (l)}$ solution at identical concentration (*squares*). No hydrogen is evolved in both cases. In the presence of nanoscrolls, only about 25 % of the $O_2$ is evolved after 5 h, indicating absorption of the added $H_2O_2$ on the nanoscrolls

**Acknowledgments** This work was supported by grant 0829142 of the National Science Foundation (NSF) and grant FG02-03ER46057 of the US Department of Energy. T. K. T. thanks NSF for a graduate student fellowship 2010 (NSFGRFP) and his group members for advice. NDB thanks the US Department of Energy for support (Grant FG02-03ER46057).

## References

1. N.S. Lewis, D.G. Nocera, P. Natl, Acad. Sci. USA **103**, 15729–15735 (2006)
2. A. Kay, I. Cesar, M. Gratzel, J. Am. Chem. Soc. **128**, 15714–15721 (2006)
3. F.A. Frame, E.C. Carroll, D.S. Larsen, M.S. Sarahan, N.D. Browning, F.E. Osterloh, *Chem. Commun.* 2206–2208 (2008)
4. F.A. Frame, F.E. Osterloh, J. Phys. Chem. C **114**, 10628–10633 (2010)
5. H. Zhou, E.M. Sabio, T.K. Townsend, T. Fan, D. Zhang, F.E. Osterloh, Chem. Mater. **22**, 3362–3368 (2010)
6. E.M. Sabio, M. Chi, N.D. Browning, F.E. Osterloh, Langmuir **26**, 7254–7261 (2010)
7. F.E. Osterloh, in *On Solar Hydrogen and Nanotechnology*, ed. by L. Vayssieres. Nanoparticle-Assembled Catalysts for Photochemical Water Splitting, ch. 16 (Wiley VCH, Weinheim, 2010)
8. O.C. Compton, F.E. Osterloh, J. Phys. Chem. C **113**, 479–485 (2009)
9. M.C. Sarahan, E.C. Carroll, M. Allen, D.S. Larsen, N.D. Browning, F.E. Osterloh, J. Solid State Chem. **181**, 1681–1686 (2008)
10. E.C. Carroll, O.C. Compton, D. Madsen, F.E. Osterloh, D.S. Larsen, J. Phys. Chem. C **112**, 2394–2403 (2008)

11. O.C. Compton, E.C. Carroll, J.Y. Kim, D.S. Larsen, F.E. Osterloh, J. Phys. Chem. C **111**, 14589–14592 (2007)
12. M.R. Allen, A. Thibert, E.M. Sabio, N.D. Browning, D.S. Larsen, F.E. Osterloh, Chem. Mater. **22**, 1220–1228 (2010)
13. M. Gasperin, M.T. Lebihan, J. Solid State Chem. **43**, 346–353 (1982)
14. K. Domen, A. Kudo, M. Shibata, A. Tanaka, K. Maruya, T. Onishi, J. Chem. Soc. Chem. Commun. **23**, 1706–1707 (1986)
15. K. Sayama, A. Tanaka, K. Domen, K. Maruya, T. Onishi, J. Phys. Chem. **95**, 1345–1348 (1991)
16. G.B. Saupe, C.C. Waraksa, H.-N. Kim, Y.J. Han, D.M. Kaschak, D.M. Skinner, T.E. Mallouk, Chem. Mater. **12**, 1556–1562 (2000)
17. R. Kaito, K. Kuroda, M. Ogawa, J. Phys. Chem. B **107**, 4043–4047 (2003)
18. K. Teshima, Y. Niina, K. Yubuta, T. Suzuki, N. Ishizawa, T. Shishido, S. Oishi, Eur. J. Inorg. Chem. **29**, 4687–4692 (2007)
19. B.Y. Xia, J.N. Wang, X.X. Wang, J. Phys. Chem. C **113**, 18115–18120 (2009)
20. M. Yagi, E. Tomita, T. Kuwabara, J. Electroanal. Chem. **579**, 83–88 (2005)
21. F. Mafune, J.Y. Kohno, Y. Takeda, T. Kondow, J. Phys. Chem. B **107**, 4218–4223 (2003)
22. P.G. Hoertz, Y.I. Kim, W.J. Youngblood, T.E. Mallouk, J. Phys. Chem. B **111**, 6845–6856 (2007)
23. G.S. Nahor, P. Hapiot, P. Neta, A. Harriman, J. Phys. Chem. **95**, 616–621 (1991)
24. A. Mills, T. Russell, J. Chem. Soc.-Faraday Trans. **87**, 1245–1250 (1991)
25. I. Burke, P. Nugent, Gold Bull. **30**, 43–47 (1997)
26. D.R. Merrill, I.C. Stefan, D.A. Scherson, J.T. Mortimer, J. Electrochem. Soc. **152**, E212–E221 (2005)
27. P. Vanysek, in *CRC Handbook of Chemistry and Physics*, 88th edn, (Internet Version 2008). (CRC Press/Taylor and Francis, Boca Raton, 2008)
28. A.J. Bard, L.R. Faulkner, *Electrochemical Methods: Fundamentals and Applications* (Wiley, New York, 2001)
29. A. Ishikawa, T. Takata, J.N. Kondo, M. Hara, H. Kobayashi, K. Domen, J. Am. Chem. Soc. **124**, 13547–13553 (2002)
30. K. Akatsuka, G. Takanashi, Y. Ebina, N. Sakai, M. Haga, T. Sasaki, J. Phys. Chem. Solids **69**, 1288–1291 (2008)
31. A. Iwase, H. Kato, A. Kudo, Chem. Lett. **34**, 946–947 (2005)

# Chapter 3
# The Oxygen Evolution Reaction: Water Oxidation Photocatalysis—Photocatalytic Water Oxidation with Suspended alpha-$Fe_2O_3$ Particles—Effects of Nanoscaling

## 3.1 Introduction

One of the most important challenges of the century will likely involve a method for converting sunlight into storable chemical fuels. Discovering a cheap and active material for solar water splitting to form hydrogen from water and sunlight could uncover an economically competitive source of renewable power. Currently however, due to low overall conversion efficiencies and photocorrosion, water splitting photocatalysts have not yet reached 10 % efficiency and long-term stabilities. Nano-scaling of bulk metal-oxide/sulfide/selenides offers new opportunities to develop more efficient catalysts, by taking advantage of increased surface area, shorter paths for charge and mass transport, and quantum confinement effects. Bulk-CdSe, for example was shown to be active for solar water reduction as nano-CdSe [1, 2]. Iron oxide ($\alpha$-$Fe_2O_3$, hematite) is an attractive candidate for these applications due to its low cost, availability, low toxicity, high photo/heat-stability, suitable band gap ($E_g = 2.06$ eV, $\lambda = 600$ nm) and valance band edge position. Despite these qualities, hematite suffers from its short hole diffusion length (204 nm) [3], short exciton lifetime ($\sim 10$ ps) [4, 5], poor minority charge carrier mobility (0.2 $cm^2\ V^{-1}\ s^{-1}$) [6], and finite light penetration depth ($\alpha^{-1} = 118$ nm at $\lambda = 550$ nm) [7], leading to a relatively poor photo-anode material. Bulk-$Fe_2O_3$ has been altered to mitigate these issues by doping with metal cations including $Mg^{2+}$, $Al^{3+}$ $Ca^{2+}$, $Ti^{4+}$, $V^{5+}$, $Cu^{2+}$, $Zn^{2+}$, $Nb^{5+}$, $Sn^{4+}$, $Pt^{4+}$ and $CO_2^{+}$ [8–13] to decrease defects. Other attempts involve selective film growth on the most active crystal face (001) for more efficient electron and hole transport [14], or enhancing purity via chemical vapor deposition and enhancing electron transport with Si-doping [15]. Many of the issues (fleeting electron/hole transport properties), however, can be addressed on the nanoscale size regime. Recently, chemical vapor deposition of $Fe(acac)_3$ led to pristine nanostructured hematite

---

This chapter appeared as: "T.-K. Townsend, et al., *Photocatalytic Water Oxidation with Suspended alpha-$Fe_2O_3$ Particles-Effects of Nanoscaling*, Energy Environ. Sci. **4**, 4270–4275 (2011)."

T. K. Townsend, *Inorganic Metal Oxide Nanocrystal Photocatalysts for Solar Fuel Generation from Water*, Springer Theses, DOI: 10.1007/978-3-319-05242-7_3,

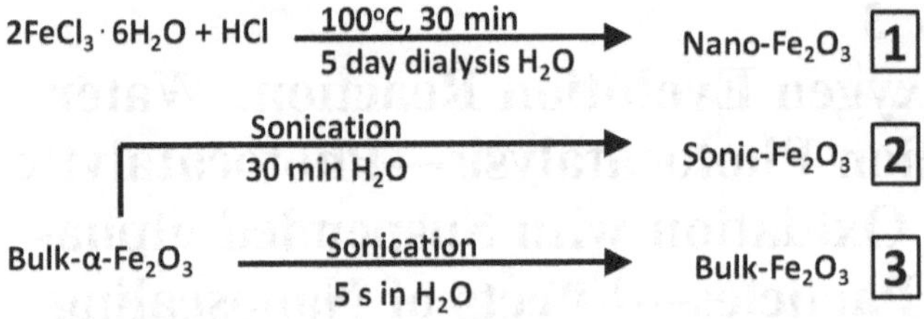

$2FeCl_3 \cdot 6H_2O + HCl$ —100°C, 30 min / 5 day dialysis $H_2O$→ Nano-$Fe_2O_3$ **1**

Sonication / 30 min $H_2O$→ Sonic-$Fe_2O_3$ **2**

Bulk-α-$Fe_2O_3$ —Sonication / 5 s in $H_2O$→ Bulk-$Fe_2O_3$ **3**

**Scheme 3.1** Bottom-up approach to give Nano-$Fe_2O_3$ **[1]** and top-down approach to prepare Sonic-$Fe_2O_3$ **[2]** from Bulk-$Fe_2O_3$ **[3]**. Sonication was performed with an ultrasonic horn at 3.65 W

films on a fluorine-doped $SnO_2$ electrode. After Si doping (added as tetraethoxysilane), $\alpha$-$Fe_2O_3$ films were shown to achieve 2.2 mA $cm^{-2}$ photocurrent densities and a IPCE of 42 % at 1.23 eV vs. RHE) [15]. In another example, $Fe_2O_3$ photoelectrodes were fabricated from $Fe_2O_3$ nanocrystals or from $Fe_3O_4$ precursor particles, followed by an annealing step [16, 17]. These electrodes showed increased water oxidation activity, but also contained Si and Pt dopants that lead to increased conductivity. To evaluate the effect of nanostructuring in the absence of doping, we report here on photocatalytic water oxidation with suspended bulk and nanoscale $\alpha$-$Fe_2O_3$ crystals. The study shows for the first time, that individually suspended $Fe_2O_3$ nanoscale particles can catalyze water oxidation in the presence of a chemical bias (+0.62 V, NHE from 1.0 mM $AgNO_3$) with a quantum efficiency of up to 0.61 % at $\lambda = 375$ nm compared to bulk-scale equivalent which was found to be not catalytic.

## 3.2 Results and Discussion

Synthesis pathways for Nano-$Fe_2O_3$ **[1]**, Sonic-$Fe_2O_3$ **[2]**, and Bulk-$Fe_2O_3$ **[3]** photocatalysts are summarized in Scheme 3.1. A bottom-up approach to producing nanospheres of $Fe_2O_3$ **[1]** followed a preparation from $FeCl_3$ and aqueous HCl precursors published by Raming et al. [18]. Alternatively, a top-down approach also produced nanoscale particles **[2]** through high intensity sonication for 30 min, and suspensions of Bulk-$Fe_2O_3$ were produced by briefly (5 s) sonicating commercial alpha $Fe_2O_3$ (hematite) in water **[3]**.

Crystallinity and phase of the particles were confirmed with powder XRD spectra as shown in Fig. 3.1. Bulk and Sonic-$Fe_2O_3$ show diffraction peaks that correspond to the alpha phase of hematite. Due to the lattice confinement the peaks of Nano-$Fe_2O_3$ display broadening and larger full width at half maximum (FWHM) including an unexpected peak at 47° signifying the presence of $\beta$-FeOOH (411 peak) [19], which is a known intermediate during aqueous synthesis of hematite [20]. Average Scherrer crystal domains sizes were calculated from the FWHM values of the (113) diffraction peak at 40° $2\theta$ to give 120 nm for Bulk-$Fe_2O_3$, 44 nm for Sonic-$Fe_2O_3$, and 5.4 nm for Nano-$Fe_2O_3$.

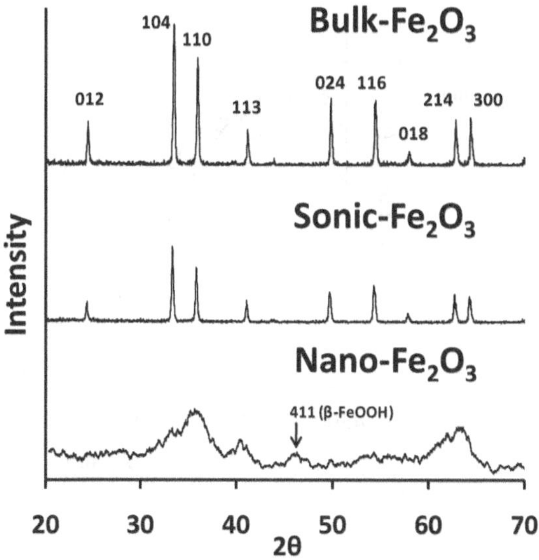


**Fig. 3.1** Powder X-ray diffraction patterns of Bulk-$Fe_2O_3$, Sonic-$Fe_2O_3$, and Nano-$Fe_2O_3$. The patterns are in agreement with the alpha structure type, with Nano-$Fe_2O_3$ also containing traces of $\beta$-FeOOH

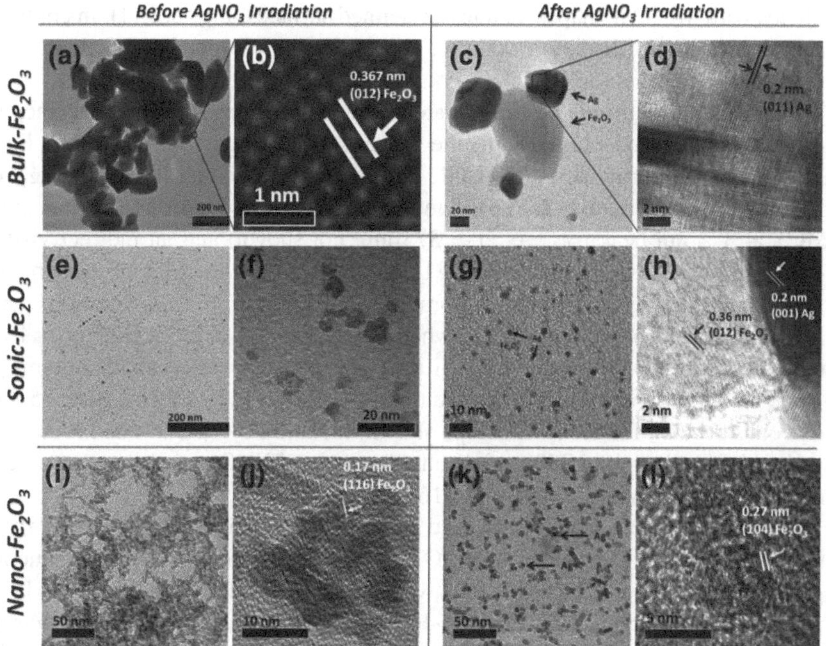


**Fig. 3.2** TEM (**a, e, f**) and HRTEM of non-irradiated α-$Fe_2O_3$ (**a, b, e, f, i, j**) and after full-spectrum $AgNO_3$ irradiation (**c, d, g, h, k, l**). Bulk-$Fe_2O_3$ (**a–d**), Sonic-$Fe_2O_3$ (**e–h**), and Nano-$Fe_2O_3$ (**i–l**). Image g is not representative of particle size. It was chosen to demonstrate silver particle formation

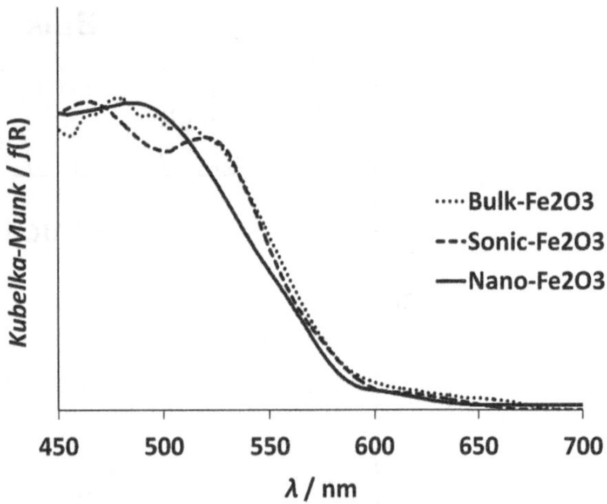


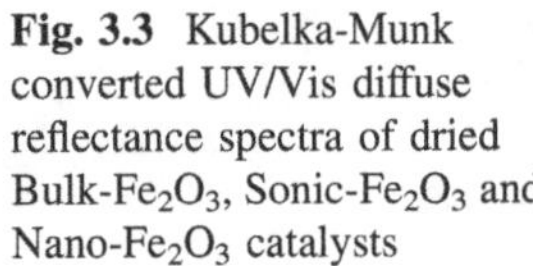
**Fig. 3.3** Kubelka-Munk converted UV/Vis diffuse reflectance spectra of dried Bulk-$Fe_2O_3$, Sonic-$Fe_2O_3$ and Nano-$Fe_2O_3$ catalysts

These Scherrer diameters are in good agreement with TEM/HRTEM images of the particles shown in Fig. 3.2. Bulk-$Fe_2O_3$ consists of large irregular and polydispersed (50–300 nm) crystals showing characteristic (001) lattice spacing of α-$Fe_2O_3$ (Fig. 3.2a, b). Comparatively, Sonic-$Fe_2O_3$ shows smaller (5–20 nm) fragmented particles with irregular shapes, and lesser defined surfaces (Fig. 3.2e, f). In contrast, Nano-$Fe_2O_3$ images display a range of single crystalline spheres (5.2 ± 1.0 nm) interspersed with rods (5.0 ± 0.8 nm × 15.2 ± 1.1 nm, Fig. 3.2i, j).

Absorbance spectra of the three materials were taken as diffuse reflectance of the dried powder and plotted with the Kubelka–Munk function (Fig. 3.3). Each material has the characteristic visible indirect transition at the band edge of 600 nm corresponding to a 2.06 eV band gap. There is also a direct UV transition LMCT at 375 nm peak ($6t_{1u} \rightarrow 2t_{2g}$, 375 nm, not shown) and an indirect visible d–d transition ($^6A_1 \rightarrow {}^4E$, 535 nm) [21]. Differences in the color of each of the materials: Bulk-$Fe_2O_3$ (light red/orange), Sonic-$Fe_2O_3$ (pink/orange), and Nano-$Fe_2O_3$ (dark blood red, orange when diluted) can also be accounted for by a slight blue shifting of this d–d transition. The presence of a small amount of β-FeOOH in the sample (a yellow precursor to α-$Fe_2O_3$ [22], $E_g$ = 2.12–2.35 eV) [19] could account for the lower slope shown in Nano-$Fe_2O_3$. Sub-band scattering tails in the region of 600–750 nm are characteristic of $Fe_2O_3$ nanostructures and is shown to increase with particle size [16].

Oxygen evolution under visible and full-spectrum light irradiation from these catalyst materials was tested with $Fe_2O_3$ powder suspensions in 1.0 mM aqueous $AgNO_3$ solutions. Stirred suspensions were irradiated with a 300 W Xe-arc lamp equipped with a long pass filter (>400 nm) or without it for full spectrum (350–800 nm). Oxygen bubble formation occurs from photogenerated holes on the surface of the photocatalyst particles (Fig. 3.5) while reduced silver metal is

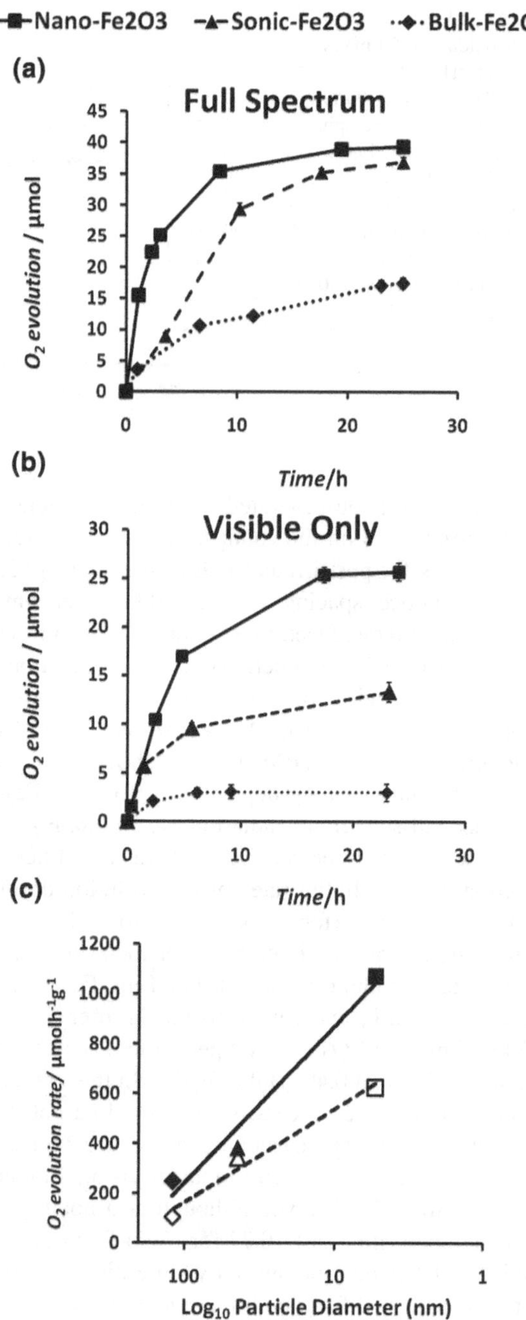


**Fig. 3.4** $O_2$ evolution from 35 μmol (5.6 mg) $\alpha$-$Fe_2O_3$ in 1.0 mM $AgNO_3$ at pH = 6.5 under full spectrum irradiation (**a**) and under visible light, $\lambda > 400$ nm (**b**). Error Bars represent deviations of two separate experiments. (**c**) Relation between initial rate of water oxidation (μmol $O_2$ $h^{-1}$ $g^{-1}$) and average Scherrer crystallite diameter (nm) derived from Fig. 3.1 in full spectrum irradiation (*solid line*) and in visible light (*broken line*) for Bulk-$Fe_2O_3$ (*diamond*), Sonic-$Fe_2O_3$ (*triangle*) and Nano-$Fe_2O_3$ (*square*)

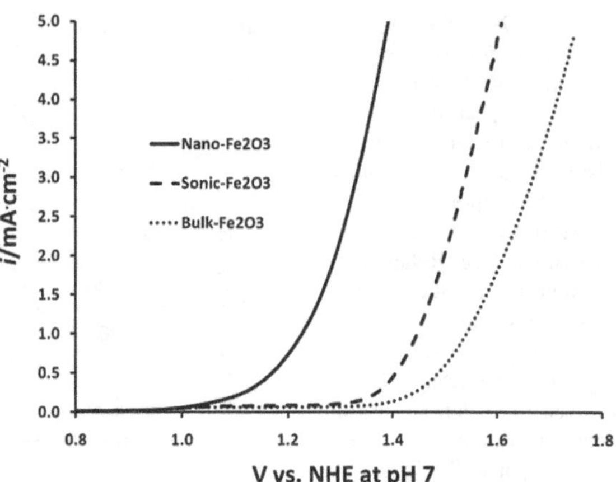


**Fig. 3.5** Electrochemical water oxidation, 50 mV/s scans, at pH = 7 in $Na_2HPO_4/NaH_2PO_4$ buffer. $Fe_2O_3$ solutions were spin coated and annealed at 450 °C on a 1 $cm^2$ gold electrode. Inset shows photo-onset potential, 10 mV/s, pH 0, 1.0 M HCl, Nano-$Fe_2O_3$ dried and annealed at 150 °C on a 1.0 $cm^2$ gold electrode

produced from photogenerated electrons. $Ag^+$ ions in solution are deposited onto electron-rich sites on the catalyst and can be observed with TEM after the reaction. Images for post-irradiated Bulk-$Fe_2O_3$ (Fig. 3.2c, d) show the 0.2 nm characteristic lattice spacing of Ag (011). Interestingly, we observe preferential deposition onto one facet, most likely the (001) facet, which can be explained by a higher anisotropic conductivity in this direction [14]. Notably, the material retained crystallinity after 24 h full spectrum irradiation (Fig. 3.2l) and no signs of photocorrosion (apart from Ag deposition). Visibly, the suspensions turn grey after irradiation, and the TEM images show large Ag particles on Sonic-$Fe_2O_3$ (Fig. 3.2h) and some Ag deposition onto Nano-$Fe_2O_3$.

Under full-spectrum irradiation (>250 nm, Fig. 3.4a), the highest photoactivity was observed from the samples and was quantified with gas chromotography as a function of time. Initial rates of $O_2$ evolution during the first 10 min range from 250 μmol $h^{-1}$ $g^{-1}$ (for Bulk-$Fe_2O_3$) to 1071 μmol $h^{-1}$ $g^{-1}$ (for Nano-$Fe_2O_3$). After this, the rate declines as silver metal deposits on the surface, blocking the active sites for photo-reactions. For Bulk-$Fe_2O_3$ this occurs after 8 h, for Sonic-$Fe_2O_3$ after 10 h, and for Nano-$Fe_2O_3$ after 5 h. After 24 h, turnover numbers (TON = moles of $O_2$ evolved per mole of $Fe_2O_3$) reach 1.13 (Nano-$Fe_2O_3$), 1.10 (Sonic-$Fe_2O_3$) and 0.49 (Bulk-$Fe_2O_3$). In this case, a turnover number above unity signifies a catalytic process, which could be potentially longer-lived in the presence of non-limiting sacrificial electron acceptors.

These materials also perform under visible irradiation (>400 nm, Fig. 3.4b) and the same trend is observed although at a notably lower rate. As a consequence, TON numbers after 24 h (0.73 for Nano-$Fe_2O_3$, 0.38 for Sonic-$Fe_2O_3$ and 0.08 for Bulk-$Fe_2O_3$) remain below unity. The higher activity under full spectral illumination is expected from both the higher absorbance of the material in the UV and from the higher oxidizing power of holes created in the valance band (see discussion of Fig. 3.6).

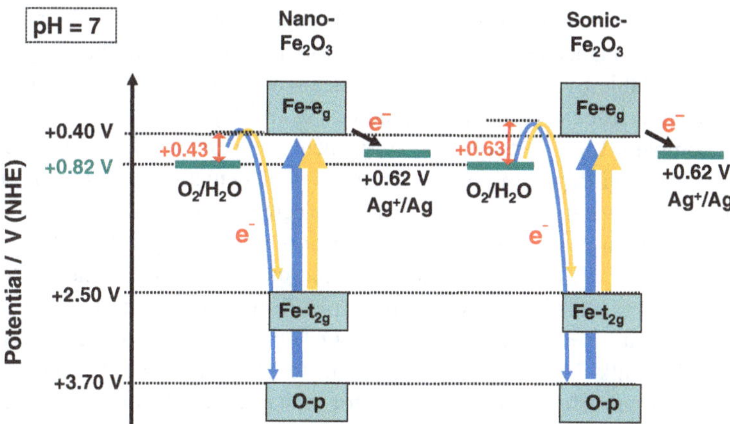


**Fig. 3.6** Energy diagram for photocatalytic water oxidation at pH = 7 with 1.0 mM $AgNO_3$ as sacrificial acceptor. The '*dark*' water oxidation overpotentials are included as an approximation for the charge transfer limitation at the $Fe_2O_3$-electrolyte interface. In reality, the anodic current measured in the dark is mediated by states in the $Fe_2O_3$ conduction band, whereas the anodic light current (*under irradiation*) involves valence band states [29, 30]

In order to visualize the effect of surface area on the initial $O_2$ evolution rates of each catalyst, initial rates were plotted against the logarithm of particle diameter (Fig. 3.4c). The rate increases inversely versus the particle diameter and is higher for full-spectrum light irradiation because of the reliance of the photoreaction on the weaker d–d transition (see discussion of Fig. 3.6). The size trend in the activity of the Nano-$Fe_2O_3$ and Sonic-$Fe_2O_3$ can be understood in terms of the short hole diffusion length (2–4 nm) [3] of hematite. Due to this inherent limitation of the material, only holes generated within a 3 nm thick sub-surface shell can participate in water oxidation. From this, one can calculate the fraction of usable e/h pairs generated in the particles as a function of size (diameters taken from powder X-ray diffraction) by comparing the volume of this shell with the total volume of the particles (assuming spheres). We calculate that the usable volume of active material is 14.3 % for Bulk-$Fe_2O_3$, 35.6 % for Sonic-$Fe_2O_3$, and 100 % for Nano-$Fe_2O_3$—since it is less than 6 nm wide. Even though this model ignores the non-spherical shape of the sonicated and bulk materials, it does explain the observed trend in oxygen evolution from each catalyst.

Quantum efficiency measurements were used to calculate the fraction of holes that actually oxidize water compared to the number of holes created within 3 nm of the surface. To do this, additional irradiations with monochromatic light from a LED (375 nm, $1.13 \times 10^7$ mol $s^{-1}$ at the flask surface), and together with observed $O_2$ generation, we calculate quantum efficiencies (QE) of Nano-$Fe_2O_3$ (0.61 %), Sonic-$Fe_2O_3$ (0.47 %), and Bulk-$Fe_2O_3$ (0.0 %) assuming four photons/holes are required to form one molecule of $O_2$. The numbers show that even in Nano-$Fe_2O_3$, >97 % of the generated charge carriers recombine before they react with water. This demonstrates that the reactivity of the nanocrystals is still limited by charge recombination (the half-life of e/h pairs in hematite is ~10 ps) [4, 5]. The problem of enhanced surface recombination in $Fe_2O_3$ is well known [23]. This may be addressable by terminating the surface with closed shell ions ($Al^{3+}$, $Sn^{4+}$, $Si^{4+}$), as several research groups have noted [9, 13, 24].

To accompany the gas evolution experiments, electrochemical measurements were conducted on thin films of the three types of hematite to further probe the differences in reactivity. We found that the electrochemical onset potentials for water oxidation vary with particle size: with onsets occurring between +1.25 and +1.40 V (NHE, 1 mA $cm^{-2}$). Based on the thermodynamic minimum water oxidation potential of +0.82 V versus NHE at pH 7, the overpotentials (measured at 1.0 mA $cm^{-2}$) are $\eta$ = +0.43 V (Nano-$Fe_2O_3$), $\eta$ = +0.63 V (Sonic-$Fe_2O_3$), and $\eta$ = +0.72 V (Bulk-$Fe_2O_3$). After adjustment to pH = 7, these data are comparable to values published in the literature for alpha-$Fe_2O_3$ films (+0.57 V [12, 15] and +0.47 V [13]). In addition to differences in particle size, Sonic-$Fe_2O_3$ and Nano-$Fe_2O_3$ have unique surface structures, which we attribute to their differences in catalytic potentials. As a product of cavitation-induced particle fragmentation [25], Sonic-$Fe_2O_3$ particles have a rough surface dominated by high index planes. This can be seen well in Fig. 3.2h. Nano-$Fe_2O_3$ on the other hand, is a product of strain-free growth, which favors the formation of stable low index facets during colloidal synthesis.

Based on the UV/Vis and electrochemical data, potential diagrams for Nano- and Bulk-$Fe_2O_3$ were drawn and shown in Fig. 3.6. The literature value of $E_{CB}$ = +0.40 V for the conduction band edge of alpha-$Fe_2O_3$ at pH = 7 [26], and the optical band gaps for the d–d and LMCT transitions were used to calculate band positions for possible catalytic electron/hole transitions. Holes for oxygen evolution can be generated either in Fe-$t_{2g}$ orbitals (visible light excitation) or in the O-p band (UV excitation). Holes in the O-p band are more oxidizing, which is a possible explanation for the greater $O_2$ evolution activity of $Fe_2O_3$ under UV excitation. The overpotentials for water oxidation are lower for Nano-$Fe_2O_3$ (+0.43 V) than for Sonic-$Fe_2O_3$ (+0.63 V), and as the diagram shows, this leads to a slightly lower energetic barrier for injecting holes into water. In addition to having more efficient hole collection at the surface (see above), this is one of the reasons for the higher activity of Nano-$Fe_2O_3$.

In the presence of silver ions, the oxygen evolution has an electrochemical bias (about +0.62 V for 1.0 M $AgNO_3$ at pH = 7), which is required for the reaction to proceed. This is slightly higher than for other reported nanostructured $\alpha$-$Fe_2O_3$ photoanodes (+0.39 V, NHE at pH 7) [15]. Based on the difference between these two potentials, the thermodynamic driving force for electron transfer from the $Fe_2O_3$ conduction band edge to the silver ion acceptor is 0.22 eV per electron. This is made possible by the fact that the silver reduction potential of 0.62 V is slightly above the water oxidation potential (0.82 V at pH = 7). Because of the small quantum yield (less than 1 %), the overall process is strongly exergonic although some conversion of light energy into chemical energy is observed (0.20 eV per converted photon).

## 3.3 Conclusion

We have demonstrated the suspensions of non-doped alpha-$Fe_2O_3$ crystals are active for water oxidation to produce catalytic amounts of $O_2$ in the presence of water and light. We observed initial $O_2$ evolution rates up to 1072 μmol $h^{-1}$ $g^{-1}$ for Nano-$Fe_2O_3$ under full-spectrum irradiation and 767 μmol $h^{-1}$ $g^{-1}$ under visible light. Catalytic rates decline after silver metal is deposited on the surfaces of the crystals, and despite this, turnover numbers above unity were achieved for nanocrystals of $Fe_2O_3$, but not for the bulk. As the particle size decreases, the catalytic activity increases as explained by the reduced hole diffusion pathway from exciton to the surface of the particle. Surface charge recombination still remains the limiting factor in these systems based on the low calculated quantum efficiencies (0.61 %) converting energy from photogenerated holes to chemical potential energy. Differences in surface structure between Nano-$Fe_2O_3$ and Sonic-$Fe_2O_3$ point to reasons for the lower water oxidation overpotentials for Nano-$Fe_2O_3$ (0.43 V at 1.0 mA $cm^{-2}$) and the favorable bottom-up synthesis. This is the first report of direct catalytic $O_2$ generation from $Fe_2O_3$ was achieved via UV and visible light irradiation in homogeneous phase. Sonication from bulk material, however, does have advantages over synthesis due to the simplicity of the preparation, and can lead to more active materials. Based on the increased photocativity of $Fe_2O_3$ from nanoscaling, both bottom-up and top-down routes may be applicable to other transition metal oxides as well.

## 3.4 Experimental

Chemicals ($\alpha$-$Fe_2O_3$ 99.9 %, $AgNO_3$99.9 %, $FeCl_3 6H_2O$ 99.9 %, and 12.1 N HCl 99.99 %) were purchased from Fisher Scientific, Pittsburg, PA. They were of reagent quality and were used as received. Water was purified to 18 MΩ cm resistivity using a Nano-pure system. Bulk-$Fe_2O_3$ was prepared with as-purchased $\alpha$-$Fe_2O_3$ and dissolved in water followed by a 5 s sonication with a Fisher Scientific FS20 ultrasonic cleaner to disperse the colloid. Sonic-$Fe_2O_3$ was prepared by ultrasonicating Bulk-$Fe_2O_3$ for 30 min with a Sonics Vibracel$^{TM}$ VCX130 ultrasonic horn at 20 % power (3.65 W) in 30 mL of $H_2O$. Nano-$Fe_2O_3$ was synthesized following a literature method [18] by dissolving $FeCl_3 6H_2O$ into 250 mL of 0.002 M HCl at 100 °C to make a 0.02 M $Fe^{3+}$ solution with stirring. After maintaining the temperature for 30 min, the heat was removed and the solution was allowed to return to room temperature. Excess spectator ions ($Fe^{3+}$, $H^+$ and $Cl^-$, etc.) were removed during a 5 day dialysis using 6–8 k MWCO tubing in 4 L water. The water was replaced daily until the conductivity reached 4.0 μS $cm^{-1}$. Full chloride removal is important during this step to avoid formation of photoactive AgCl during irradiation [27]. The final suspension contained 1.12 mg $mL^{-1}$ of $Fe_2O_3$ as determined by gravimetric analysis.

Transmission electron images were taken with a Philips CM-12 instrument at 120 kV acceleration potential. Bright field high resolution transmission electron microscopy (HRTEM) images were taken using a JEOL 2500SE 200 kV TEM. Copper grids with a carbon film were dropped into aqueous dispersions of the samples followed by washing with water and air drying.

UV/Vis diffuse reflectance spectra were taken as dried powders on white Teflon tape using an Ocean Optics HR2000 CG-UVNIR spectrometer and a DH2000 light source. Reflectance was converted to Kubelka–Munk as $f(R) = (1\text{-}R)^2(2R)^{-1}$ vs. wavelength to correct for scattering.

For electrochemical measurements thins films of the catalysts were prepared on a gold foil electrode (1.0 $cm^2$) by drop coating and annealing at 450 °C. The films were then pressed with a CARVER 4350.L steel press to at 15000 psi following literature procedures [28]. A wire was attached to the bare gold back with conductive carbon tape and sealed with adhesive. The electrode was placed into a $N_2$-purged 3-electrode cell with a Pt counter electrode and a saturated calomel reference electrode connected to the cell with a KCl salt bridge. The cell was filled to 50 mL with 0.25 M $Na_2HPO_4/NaH_2PO_4$ buffer solution at pH 7 with a constant stream of $N_2$ above the solution. Dark cyclic voltammetry scans were taken at 50 mV/s. The system was calibrated using the redox potential of $K_4[Fe(CN)_6]$ at +0.358 V (NHE).

The rate of photochemical oxygen evolution from each catalyst was determined by irradiating 5.6 mg (35 μmol) of $Fe_2O_3$ in 1.0 mM $AgNO_3$ in 50 mL of water in a quartz round bottom flask with a 300 W Xe arc lamp (1.4 W $cm^{-2}$ at flask, $\lambda = 250$–$680$ nm), measured with a UV/Vis GaAsP photodetector. Visible light ($\lambda > 400$ nm) was achieved by inserting a 400 nm long-pass filter between the light and the flask. The air-tight irradiation system connects a vacuum pump and a gas chromatograph (Varian 3800) with the sample flask to quantify the amount of gas evolved, using area counts of the peaks and the identity of the gas from the calibrated carrier times. Prior to irradiation, the flask was evacuated down to 5 torr and purged with argon gas. This cycle was repeated until the chromatogram of the atmosphere above the solution indicated that the sample did not contain hydrogen, oxygen, or nitrogen. A solution of 1.0 mM $AgNO_3$ was tested as a control under full spectrum irradiation, and less than 3 μmol of $O_2$ were evolved after 10 h. Quantum efficiency measurements were conducted with a 375 nm LED (3.42 mW $cm^{-2}$, $1.13 \times 10^{17}$ photon $s^{-1}$) in place of the lamps.

Powder XRD scans were conducted with a Scintag XRD, $\lambda = 0.154$ nm at −45 kV and 40 mA with tube slit divergence (2 mm), scatter (4 mm), column scatter (0.5 mm), and receiving (0.2 mm). The FWHM of the peaks were calculated with ThemoARL DMSNT software (version 1.39–1).

**Acknowledgments** FEO thanks Research Corporation for Science Advancement for a Scialog award. This work was further supported by the National Science Foundation (NSF, grant 0829142) and by the US Department of Energy (grant FG02–03ER46057). TKT thanks NSF for a Graduate Research Fellowship 2011. N.D.B. thanks the US Department of Energy for support (grant FG02–03ER46057).

## References


1. F.A. Frame, E.C. Carroll, D.S. Larsen, M. Sarahan, N.D. Browning, F.E. Osterloh, Chem. Commun. **19**, 2206–2208 (2008)
2. F.A. Frame, F.E. Osterloh, J. Phys. Chem. C **114**, 10628–10633 (2010)
3. J.H. Kennedy, K.W. Frese, J. Electrochem. Soc. **124**, C130–C130 (1977)
4. A.G. Joly, J.R. Williams, S.A. Chambers, G. Xiong, W.P. Hess, D.M. Laman, J. Appl. Phys. **99**, 6 (2006)
5. N.J. Cherepy, D.B. Liston, J.A. Lovejoy, H.M. Deng, J.Z. Zhang, J. Phys. Chem. B **102**, 770–776 (1998)
6. A.J. Bosman, H.J. Vandaal, Adv. Phys. **19**, 1–7 (1970)
7. K. Itoh, J.O. Bockris, J. Electrochem. Soc. **131**, 1266–1271 (1984)
8. S. Mohanty, J. Ghose, J. Phys. Chem. Solids **53**, 81–91 (1992)
9. J.A. Glasscock, P.R.F. Barnes, I.C. Plumb, N. Savvides, J. Phys. Chem. C **111**, 16477–16488 (2007)
10. T. Arai, Y. Konishi, Y. Iwasaki, H. Sugihara, K. Sayama, J. Comb. Chem. **9**, 574–581 (2007)
11. C.J. Sartoretti, B.D. Alexander, R. Solarska, W.A. Rutkowska, J. Augustynski, R. Cerny, J. Phys. Chem. B **109**, 13685–13692 (2005)
12. Y.S. Hu, A. Kleiman-Shwarsctein, A.J. Forman, D. Hazen, J.N. Park, E.W. McFarland, Chem. Mater. **20**, 3803–3805 (2008)
13. R.L. Spray, K.J. McDonald, K.S. Choi, J. Phys. Chem. C **115**, 3497–3506 (2011)
14. C.M. Eggleston, A.J.A. Shankle, A.J. Moyer, I. Cesar, M. Gratzel, Aquat. Sci. **71**, 151–159 (2009)
15. A. Kay, I. Cesar, M. Gratzel, J. Am. Chem. Soc. **128**, 15714–15721 (2006)
16. K. Sivula, R. Zboril, F. Le Formal, R. Robert, A. Weidenkaff, J. Tucek, J. Frydrych and M. Gratzel, J. Am. Chem. Soc **132**, 7436–7444 (2010)
17. R.H. Goncalves, B.H.R. Lima, E.R. Leite, J. Am. Chem. Soc. **133**, 6012–6019 (2011)
18. T.P. Raming, A.J.A. Winnubst, C.M. van Kats, A.P. Philipse, J. Colloid Interface Sci. **249**, 346–350 (2002)
19. Y. Xiong, Y. Xie, S.W. Chen, Z.Q. Li, Chemistry-a Eur. J **9**, 4991–4996 (2003)
20. R.S. Sapieszko, R.C. Patel, E. Matijevic, J. Phys. Chem. **81**, 1061–1068 (1977)
21. L.A. Marusak, R. Messier, W.B. White, J. Phys. Chem. Solids **41**, 981–984 (1980)
22. E. Matijevic, P. Scheiner, J. Colloid Interface Sci. **63**, 509–524 (1978)
23. M.P. Dareedwards, J.B. Goodenough, A. Hamnett, P.R. Trevellick, J. Chem. Soc. Farad. T. I. **79**, 2027–2041 (1983)
24. F. Le Formal, N. Tetreault, M. Cornuz, T. Moehl, M. Gratzel, K. Sivula, Chemical Science **2**, 737–743 (2011)
25. K.S. Suslick, G.J. Price, Annu. Rev. Mater. Sci. **29**, 295–326 (1999)
26. Y. Xu, M.A.A. Schoonen, Am. Mineral. **85**, 543–556 (2000)
27. M. Lanz, D. Schurch, G. Calzaferri, J. Photochem. Photobiol., A **120**, 105–117 (1999)
28. D.A. Totir, B.D. Cahan, D.A. Scherson, Electrochim. Acta **45**, 161–166 (1999)
29. H. Gerischer, Semiconductor electrode reactions in *Adv. Electrochem. & Electrochem. Eng.*, ed. by P. Delahay, (Interscience, New York, 1961), **1**, pp. 139–232
30. A.J. Nozik, R. Memming, J. Phys. Chem. **100**, 13061–13078 (1996)

## References

# Chapter 4
# Overall Photocatalytic Water Splitting with Suspended NiO-$SrTiO_3$ Nanocrystals

## 4.1 Introduction

The direct conversion of sunlight into high purity chemical fuel (hydrogen) can be accomplished through photocatalytic water splitting reactions. This process is expected to be the most promising solution to the global energy dilemma based on a recent DOE report where suspended heterogeneous photocatalysts operating at 10 % solar energy conversion efficiency would be theoretically able to produce hydrogen fuel at a cost of \$1.63/kg $H_2$ [1]. Considering the cost of fabrication and installation of photoelectrochemical (PEC) cells or from the combination of photovoltaic (PV) cells with water electrolysers, hydrogen produced from suspended photocatalysts is considerably cheaper. Materials of choice for this process include inorganic nanomaterials. Due to their small size, full light penetration can be achieved, and even short-lived charge carriers can reach the semiconductor-liquid interface. Nanoscaling has proven to be beneficial for catalytic activity for water oxidation catalyts: $IrO_2$ [2] $Fe_2O_3$ [3–7] and $MnO_2$ [8–10], all of which have been shown to photooxidize water in the presence of an electrical or chemical bias. One disadvantage of nanomaterials is that on this size scale, space charge layers are not effective for separating electron hole pairs, and recombination is enhanced in the absence of a bias. This is one of the reasons that nanoscale catalysts for *overall water splitting* are very rare. So far only three systems have been reported in the literature, all of which require deep ultraviolet light ($\lambda < 270$ nm) for operation. Kudo's group showed that $LiNbO_3$ nanowires (70 nm × 10 μm) can split water after modification with $RuO_2$ cocatalyst particles and the quantum yield was estimated as 0.7 % at 254 nm [11]. Yan et al. reported overall water splitting with $RuO_2$-modified $Zn_2GeO_4$ nanorods (100 × 150 nm) under UV light from a 400 W Hg UV lamp to excite the large band gap (Eg > 4.5 eV) of the material [12]. Last year, Domen's group reported overall water splitting with aggregates of NiO-loaded $NaTaO_3$ nanocrystals (20 and 40 nm), but ultraviolet light

This chapter appeared as: "T.-K. Townsend, et al., *Overall Photocatalytic Water Splitting with NiOx-$SrTiO_3$—A Revised Mechanism*, Energy Environ. Sci. **5**, 9543–9550 (2012)."

T. K. Townsend, *Inorganic Metal Oxide Nanocrystal Photocatalysts for Solar Fuel Generation from Water*, Springer Theses, DOI: 10.1007/978-3-319-05242-7_4,

($\lambda < 200$ nm) was necessary for catalyst operation [13]. These proof of concept studies are encouraging for the process of developing an efficient and stable water splitter; however significant improvements need to be realized for commercial application. For these systems to be useful for a true solar energy to fuel conversion system, a better understanding of charge separation on the nanoscale must be attained including information about the energetics and kinetics of water conversion from these materials.

In this study we describe an investigation of the properties of the first nanoscale water splitting catalyst based on NiO-modified $SrTiO_3$ (STO). Considering its lower energy band gap compared to previously studied overall water splitters, STO should allow considerably better light absorption ($h\nu > 3.2$ eV). STO is also among the best studied photocatalyst materials in the literature [14–27], and after NiO attachment, the heterogeneous catalyst has been shown to split water at high rates up to 41/20 μmol $h^{-1}$ $g^{-1}$ [18] while showing quantum efficiencies of up to 5.2 % at (420 nm) after doping with Rh and using a Pt cocatalyst [20]. In contrast, the NiO-STO system does not require rare elements for function, and we find that a 30 nm NiO-STO catalyst can split water with an efficiency comparable to the bulk material (QE estimated at 2.38 % at $\lambda = 315$ nm), but that at smaller sizes, the activity becomes increasingly limited by quantum size effects and slow water oxidation kinetics. These findings establish a lower size limit for photocatalysts based on $SrTiO_3$, and suggest that improvements of the activity should target water oxidation kinetics and charge separation in this system.

## 4.2 Results and Discussion

For comparison in this study, three types of STO were synthesized using a high temperature solid-state reaction to form Bulk-STO (>100 nm), Nano (30 nm) and Nano (6.5 nm) which is summarized in Scheme 4.1. Bulk STO was prepared from a solid-state reaction of $TiO_2$ and $SrCO_3$ precursors, and from the TEM (Fig. 4.1a), the product forms 0.1–1.0 μm particles with ill-defined surface features and of likely polycrystalline structure [28]. Alternatively, nanoscale STO particles were synthesized via a hydrothermal pressure reaction of $TiO_2$, $Sr(OH)_2$ and KOH [29] to form 25–60 nm particles with an average of 30 nm consisting mostly of rigid cubic shapes. Finally, Nano (6.5 nm) STO was prepared by exposing the bimetallic precursor ($SrTi\text{-}[OCH_2CH(CH_3)OCH_3]_6$; 0.7 M in *n*-butanol/3-methoxy-propanol) to water vapor under controlled conditions [30]. The bimetalic complex slowly oxidizes to form 6.0 + 1.0 nm STO crystals which are agglomerated within an organic matrix. Based on powder XRD diffraction patterns, each of the three catalysts are crystalline and phase identified to STO (Fig. 4.2). Using the Scherrer equation for 30 nm STO and 6.5 nm STO, crystal domain sizes were estimated from the diffraction peak widths (at half maximum), and the calculated values of 30 and 6.5 nm agree well with the TEM images of these crystals.

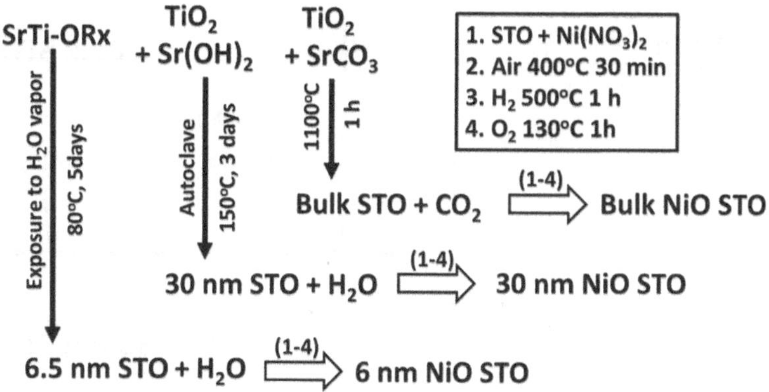


**Scheme 4.1** Synthesis of Bulk NiO STO, 30 nm NiO STO and 6.5 nm NiO STO with reduction/oxidation pre-treatment to the activate NiOx ($0 \leq x \leq 1$) cocatalyst

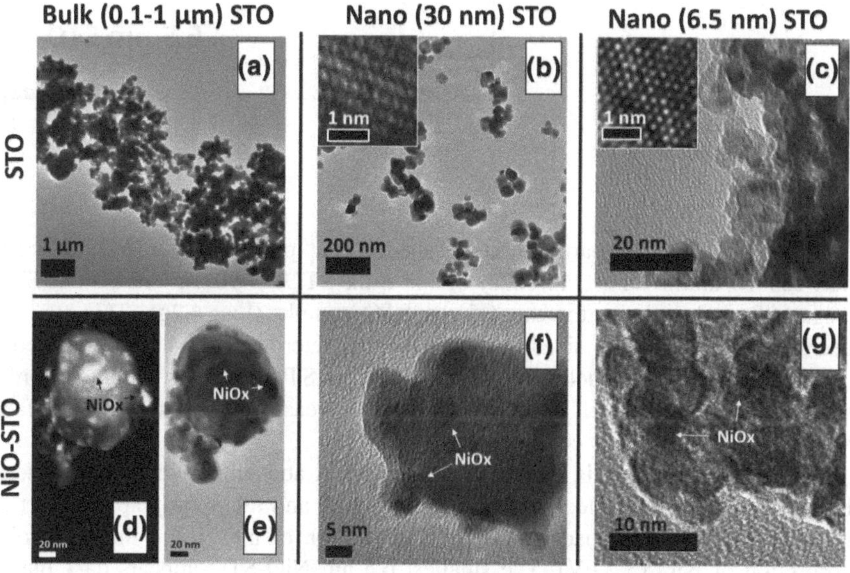


**Fig. 4.1** HRTEM of Bulk STO (**a**), 30 nm STO (**b**), 6.5 nm STO (**c**), HRSTEM (**d**) and HRTEM (**e**) of Bulk NiO-STO, HRTEM of 30 nm NiO-STO (**f**), and 6.5 nm NiO-STO (**g**)

In order to activate these semiconductor framework materials for overall water splitting, NiO was deposited onto the surface according to Scheme 4.1. Starting with $Ni(NO_3)_2$, which decomposes to Ni metal on STO, the precursors were calcined in air (400 °C), followed by reduction in hydrogen (500 °C) and then oxygen atmosphere (130 °C) to produce NiO decorated STO catalyst materials. TEM images of Bulk NiO-STO, 30 nm NiO-STO, and 6.5 nm NiO-STO are shown in Fig. 4.1. Z-contrast imaging for Bulk STO verified the presence of $NiO_x$

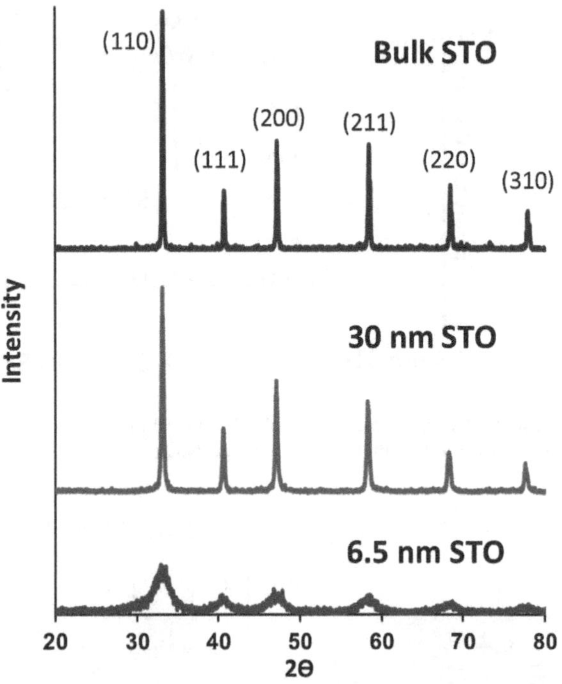


**Fig. 4.2** Powder X-ray diffraction patterns of Bulk STO, 30 nm STO, and 6.5 nm STO

($0 \leq x \leq 1$) cocatalyst particles, and embedded $Ni^0$ particles. The $NiO_x$ sizes are between 10 and 50 nm (Fig. 4.2d, e) for Bulk STO, 5–10 nm for 30 nm STO (Fig. 4.2f) and 1–4 nm for the 6.5 nm STO sample (Fig. 4.2g). This type of decrease in NiO particle sizes is expected from heterogeneous nucleation theory due to the increase in nucleation sites for smaller STO particles. In other words, smaller NiO particles will be the result of an increased area of the STO surface after of the deposition is complete.

Optically, the materials change color after NiOx attachment which can be seen in the UVVis diffuse reflectance spectra (Fig. 4.3) and the photographs (insert in Fig. 4.3a) due to the broad visible light absorption of nickel and the ultraviolet absorption of NiO (3.2 eV). The smallest 6.5 nm NiO-STO appears dark black, 30 nm NiO-STO dark grey and Bulk NiO-STO appears light grey (Fig. 4.3 photos). In contrast, the Ni-free materials are white powders with no absorption in the visible. Due to an indirect transition between the O 2p valence band edge and the Ti 3d conduction band edge, Bulk STO has an absorption onset at 400 nm. It also has a direct transition involving O 2p and Sr 4d orbitals, which is found to occur at 3.30 eV (Fig. 4.3c) [31]. The direct and indirect transitions of both 30 nm STO and 6.5 nm STO are blue shifted to higher energies (3.3–3.7 eV) compared to the bulk form. Based on the Bohr-exciton radius of SrTiO3 (6.25 nm) [32], this indicates an incipient quantum size effect for the smaller particles. The increase of band gap of 6.5 nm STO is partially obscured by the absorbance tail in the visible

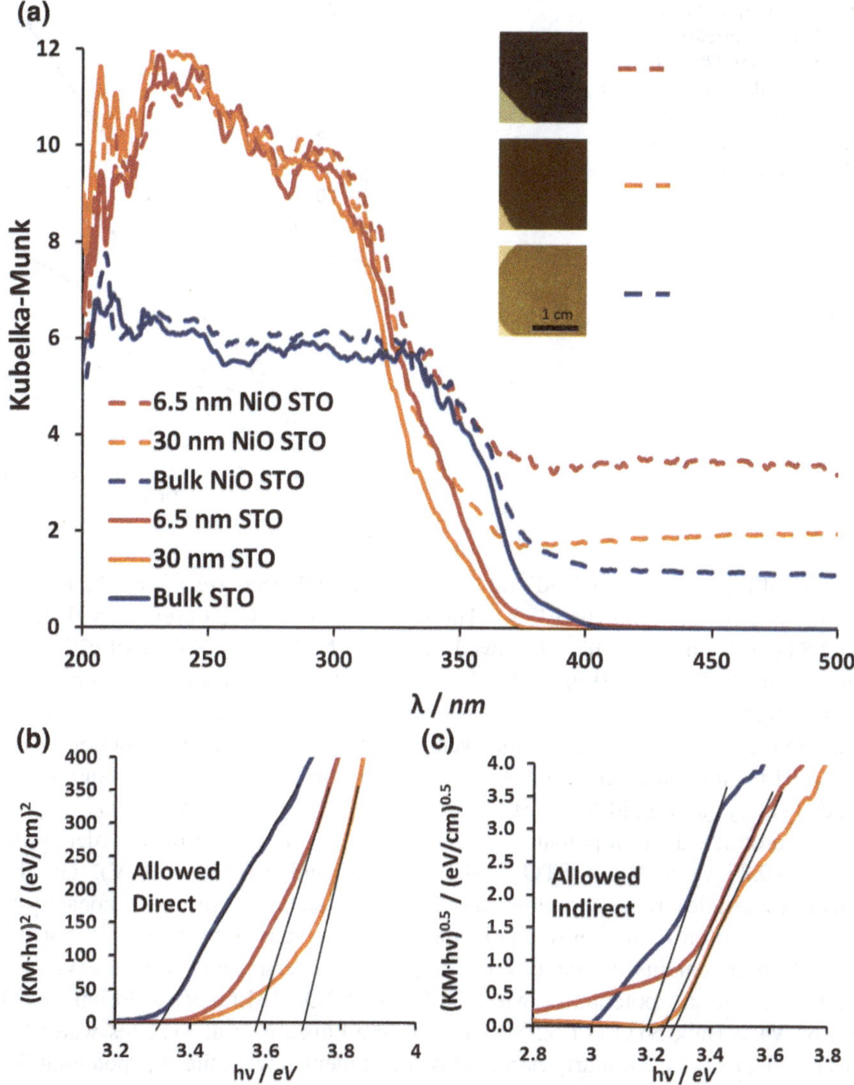


**Fig. 4.3** Kubelka-Munk diffuse reflectance of unmodified STO powders and NiOx attached STO catalysts (**a**). Tauc plot for allowed direct transitions (**b**) and allowed indirect transitions (**c**)

region (see Fig. 4.3c). This could be a result of the greater specific surface area (due to the smaller crystal size) and the introduction of mid gap defect states that are favored by the lower synthesis temperature of this material (80 °C), compared to the other syntheses (150 and 1100 °C).

These three compounds were tested for photocatalytic activity by suspending equal amounts (50 mg) in pure water and irradiated with a 300 W Xe arc lamp. In each case, stoichiometric evolution of both $H_2$ and $O_2$ was observed (Fig. 4.4).

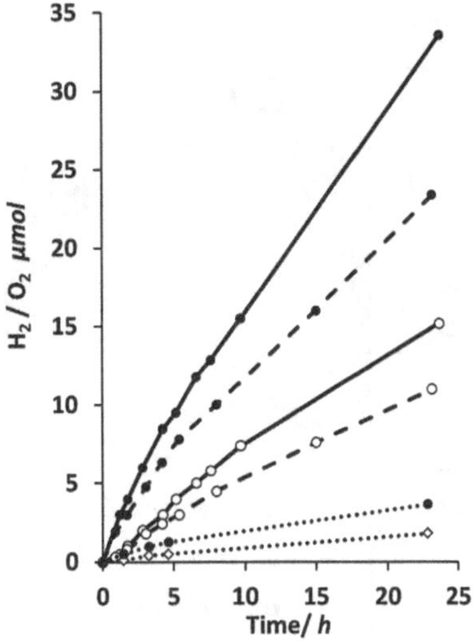


**Fig. 4.4** $H_2$ (*filled circle*) and $O_2$ (*open circle*) evolution from 272 µmol (50 mg) Bulk NiO STO (*solid line*), 30 nm NiO STO (*dashed line*) and 6.5 nm STO (*dotted line*) in 50 mL water at pH = 7 under full spectrum irradiation

Water splitting rates remained stable after an initial decrease after 5 h, for the duration of the experiments (23 h). The highest activity was observed with Bulk NiO STO (28 µmol $g^{-1}$ $h^{-1}$) followed by 30 nm NiO STO (19.4 µmol $g^{-1}$ $h^{-1}$), and 6.5 nm NiO STO (3.0 µmol $g^{-1}$ $h^{-1}$) (Fig. 4.4). Further testing was conducted to investigate reasons for decreasing activity with crystal size via electrochemical measurements. Overpotentials for water oxidation and water reduction were recorded from photoelectrochemical scans from films of the $NiO_x$-free and $NiO_x$-loaded catalysts on gold foil in 0.1 M KCl solution at pH = 7 (Fig. 4.5).

The proton reduction potentials were found occur in the following order: Bulk STO (−0.85 V) > 30 nm STO (−0.81 V) > 6.5 nm STO (−0.80 V), i.e. the smallest particles reduce protons the easiest. Attachment of $NiO_x$ cocatalysts moves all proton reduction overpotentials to lower values (by 6–9 mV), but the overall order remains the same. Interestingly, the reverse trend was observed for water oxidation potentials with: 6.5 nm STO (+1.66 V) > 30 nm STO (+1.64 V) > Bulk STO (+1.62 V), i.e. here the bulk material oxidizes water the easiest (Table 4.1). Similarly here, NiOx attachment lowers the overpotential for $O_2$ evolution, while it does not change the relative order of reactivity between catalysts. At present, the reason for the variation of redox potentials with particle size is unclear. One explanation could be an increase of the $SrTiO_3$ surface acidity with decreasing particle size, which is true for nanoscale $TiO_2$ [33]. Lower surface pH would translate into higher proton coverage, and subsequently lower proton reduction potential and also higher water oxidation potential.

The crystal sizes of the catalysts compared to the water oxidation overpotential ($\eta_{Ox}$) and also with the $H_2/O_2$ evolution rates (R) in a plot shown in Fig. 4.6. The evident correlation between $\eta_{Ox}$ and R suggests that the catalyst performance is limited by the water oxidation kinetics under illumination. We previously observed that the photocatalytic rates in nanomaterials can be correlated with the

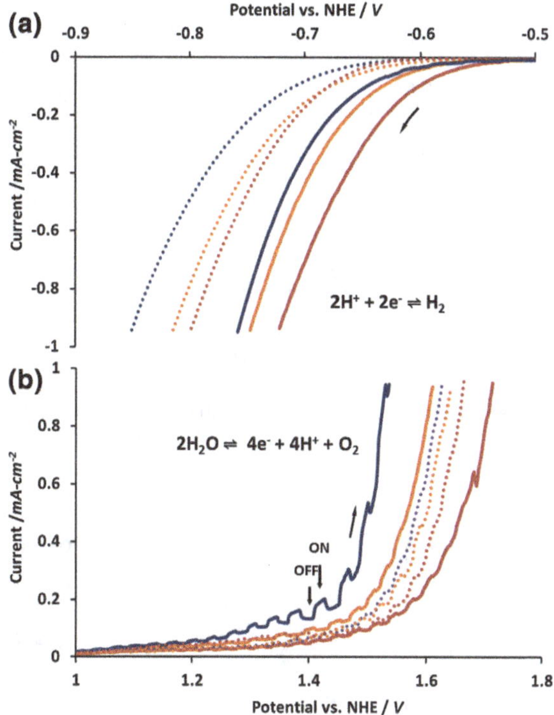


**Fig. 4.5** Electrochemical scans (10 mV $s^{-1}$) on $NiO_x$-STO films on gold in 0.25 M $Na_2HPO_4/NaH_2PO_4$ buffer (pH = 7). All potentials are versus NHE. Chopped light during the anodic sans (**b**) reveals small photocurrents. *Dotted blue* bulk STO, *solid blue* bulk NiO-STO, *dotted orange* 30 nm STO, *solid orange* 30 nm NiO STO, *dotted red* 6.5 nm STO, *solid red* 6.5 nm NiO STO

**Table 4.1** Electrochemical water reduction and oxidation potentials (V vs. NHE at 0.90 mA $cm^{-2}$)

| Catalyst | Reduction potential | Oxidation potential |
|---|---|---|
| Bulk STO | −0.85 | +1.62 |
| 30 nm STO | −0.81 | +1.64 |
| 6.5 nm STO | −0.80 | +1.66 |
| Bulk NiO STO | −0.76 | +1.53 |
| 30 nm NiO STO | −0.75 | +1.61 |
| 6.5 nm NiO STO | −0.72 | +1.71 |

Overpotentials are calculated based on the theoretical water reduction and oxidation potentials at pH 7 of −0.413 V and +0.817 V (NHE), respectively

electrochemical overpotentials [3, 34, 35]. When considering that the photooxidation process is driven by the photoholes in the valence band, whereas the electro-oxidation is driven by empty states in the conduction band, the correlation suggests that both processes have similar reaction intermediates.

Besides the kinetics, the photoenergetics of the three $SrTiO_3$ materials were also investigated, using photoelectrochemical scans on films in aqueous 0.1 M KCl solution that also contained 10 % (vol) methanol (Fig. 4.7). Because $SrTiO_3$ is intrinsically n-type, the photoonset potential can be used to approximate the conduction band edge. Onset potentials could be observed even though photocurrents

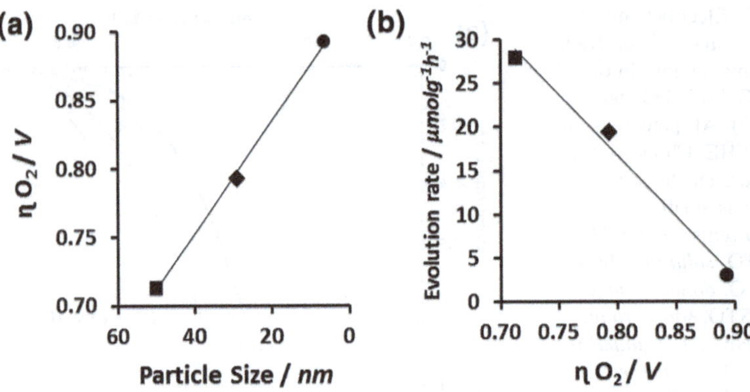


**Fig. 4.6** **a** Water oxidation overpotential vesus particle size. **b** Water oxidation potential versus gas evolution rate. *Squares* bulk NiO STO, *diamonds* 30 nm NiO STO, *circles* 6.5 nm NiO STO

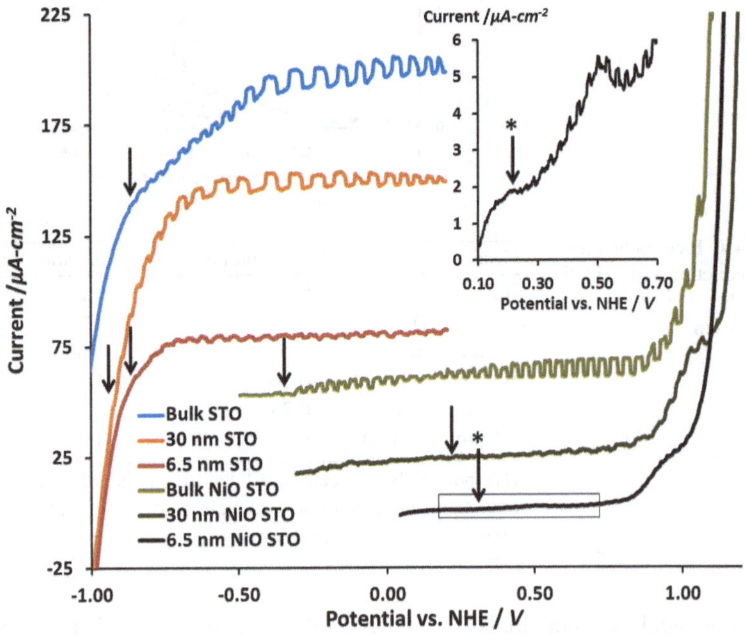


**Fig. 4.7** Photo-current scans of Bulk STO (*blue*), 30 nm STO (*orange*), 6.5 nm STO (*red*), Bulk NiO STO (*brown*), 30 nm NiO STO (*grey*) and 6.5 nm NiO STO (*black*) films on Au substrate, 10 mV s$^{-1}$, at pH = 7 in 0.1 M KCl/10 % vol. methanol

were found to be small (1.0–10 μA cm$^{-2}$), and limited by high electrical resistance within the films. As a reference, experimental potential of Bulk STO (–0.83 V vs. NHE) compares well with the literature value for the conduction band

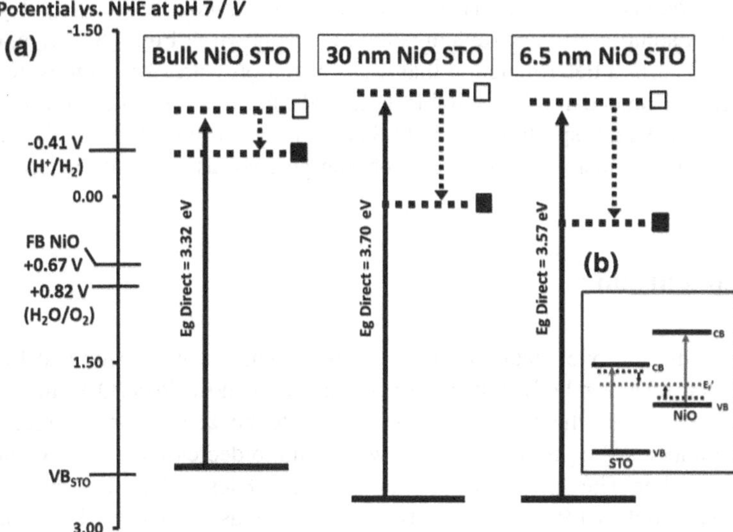


**Fig. 4.8** **a** Band edge potentials of STO before and after addition of NiO. **b** Proposed STO-NiO z-scheme for photocatalytic water splitting

edge [36]. For 30 nm STO the onset potential value is shifted in the negative direction (–0.94 V) and for 6.5 nm STO, a value similar to that of Bulk STO is found (−0.83 V). The photocurrent for this material is very weak, possibly obscuring a more negative value that would be closer to that of 30 nm STO.

Besides the optical properties, a second manifestation of the quantum size effect would be the negative shift of the conduction band edge. NiOx attachment to the catalyst surfaces causes a positive shift in photoonset potentials in each case. The strongest shift (+1.08 V) occurs for 6.5 nm STO, followed by 30 nm STO (+1.04), and finally Bulk STO (+0.50 V). As a result of this shift, the CB edge position of the 6.5 nm NiO STO composite is found at +0.25 V, closer to the Fermi level of p-NiO, for which Nakaoka et al. report a value +0.67 V (NHE) and a band gap of 3.31 eV [37]. As predicted, this observation suggests that there is Fermi level equilibration between STO and NiO when in contact. The dependence of this effect on STO particle size points to the importance of the interfacial area between NiO and STO particles. In addition, the electronic structure of the nanoscale STO particles is expected to respond more sensitively to surface effects than Bulk STO since the fraction of surface atoms inversely correlates with particle size. This is also observed in nanoscale niobates and titanates after binding of Pt and IrOx nano-cocatalysts (showing 60–240 mV positive shifts) [34, 38]. Band edge potentials of each STO catalyst (vs. NHE at pH = 7) are shown in Fig. 4.8. The most striking features are the increase of band gap energy in going to the smaller particles, and the significant shift of the conduction band edge in all samples upon adding NiO. Based on the position of $E_{CB}$, the NiO-STO samples should no longer be able to reduce protons at pH = 7. However, this contradicts the $H_2/O_2$ data in Fig. 4.4. A solution to this problem is the model shown in Fig. 4.8b. Here, n-STO and p-NiO form a heterojunction that after excitation with two photons generates one electron hole pair with suitable potentials for overall water splitting. This implies that water splitting on NiO-STO follows a four-photon Tandem (or z-scheme) mechanism, and not a two-photon process, as generally assumed [39].

## 4.3 Conclusion

In this study, the photocatalytic water splitting reaction was examined for NiO-$SrTiO_3$ particles for both bulk-STO and for two nanoscale-STO semiconductor crystals. While all the materials were shown to be active for complete water splitting under UV light, the reactivity was found to decrease as the STO particles became smaller. The 30 nm particles are about 35 % less active than the bulk, and 6.5 nm have only 10 % of the activity, and the decrease of activity is attributed to a combination of factors. The most important ones include reduced light absorption due to incipient quantum size effects observed for the nanomaterials in addition to increased water oxidation potentials with the smaller particles. This sets a lower limit for the particle size of STO above 13 nm (two times the Bohr radius), and suggests the need for additional water oxidation co-catalysts as this reaction was found to be rate limiting. The energetics of the NiO-STO systems (Fig. 4.8) provide support for a four photon tandem (or z-scheme) mechanism, instead of the generally accepted two-photon process. Studies on charge separation at the NiO-STO interface are currently underway. Lastly, we have demonstrated the first nanoscale water splitter based on NiO-$SrTiO_3$. Among the other known systems (see introduction), this one features the lowest band gap, and thus supports greatest light absorption in the UV.

## 4.4 Experimental

Chemicals (KOH 99.9 %, $Sr(NO_3)_2$ 99.9 %, P25 $TiO_2$, $NH_4OH$ 99.9 %, $Ni(NO_3)_2(H_2O)_6$ 99.9 %, ammonium hydroxide were purchased from Fisher Scientific, Pittsburgh, PA. Strontium titanium bimetallic alkoxide (SrTi-$[OCH_2CH(CH_3)OCH_3]_6$; 0.7 M in n-butanol/3-methoxypropanol; Gelest, Inc.) was used as received. $Sr(OH)_2$ was prepared by precipitation at room temperature from $Sr(NO_3)_2$ and KOH in water. They were of reagent quality and were used as received. Water was purified to 18 MΩ cm resistivity using a *Nano-pure* system.

**Bulk-$SrTiO_3$ particles** were synthesized via a high temperature solid state process [28] where 0.02 mol, 1.74 g P25 $TiO_2$ was mixed and sonicated for 10 min with 0.026 mol, 5.53 g $Sr(NO_3)_2$ in a 1:1.2 molar ratio in 200 mL of water to produce precursor for 4 g of $SrTiO_3$. Aqueous oxalic acid (0.4 M, 1:1 molar ratio with Sr) was added drop wise to the solution under vigorous stirring. Conc. ammonium hydroxide was added slowly to increase the pH of the solution to 6.5 in order to precipitate strontium oxalate crystals onto the 25 nm $TiO_2$ particle surfaces via heterogeneous nucleation [40]. When precipitation was complete, the white solid was centrifuged and washed 8 times in 50 mL of water, followed by drying in air at 100 °C. This precursor was calcined in a Thermolyne 79300 Tube Furnace to 1100 °C for 1 h with a heating rate of 10 °C $min^{-1}$. After cooling to room temperature, the white solid was washed twice in 50 mL of 5 M $HNO_3$ to remove excess $SrCO_3$ followed by repeated water washes until the supernatant reached a pH of 7. Phase purity was confirmed by powder x-ray diffraction.

**Nano-$SrTiO_3$ particles** (30 nm) were synthesized via a hydrothermal autoclave route [29]. 0.18 g, 2.3 mmol of P25 $TiO_2$ was mixed in an 20 mL aqueous solution of 1.26 g, 23 mmol of KOH and 0.508 g, 2.3 mmol of $Sr(OH)_{2\ (s)}$ in a 30 mL Teflon container sealed in a pressure vessel. This mixture was heated to 150 °C for 3 days. Afterwards, the white powder were washed with water until the supernatant reached a pH of 7 and then dried at 100 °C in air.

**Nano-$SrTiO_3$ particles** (6.5 nm) were synthesized following a vapor diffusion sol-gel technique [30] where 10 mL of a strontium titanium bimetallic alkoxide: $SrTi\text{-}[OCH_2CH(CH_3)OCH_3]_6$; 0.7 M in *n*-butanol/3-methoxypropanol; Gelest, Inc. was heated under $N_2$ atmosphere at 80 °C. This heated solution was exposed to water vapor from an adjacent flask of water (at 25 °C) for 72 h. The resulting off-white gel was washed in ethanol and dried at 100 °C to yield 6.5 nm STO particles.

**NiO attachment-Reduction/Oxidation** The $SrTiO_3$ particles (200 mg, 1.10 mmol) were added to an aqueous $Ni(NO_3)\cdot 6H_2O$ solution (0.023 g, 3 wt % loading of NiO:STO) and thoroughly mixed in a sonication bath for 10 min. This solution was dried at 100 °C and then annealed in air for 30 min at 400 °C and form NiO islands on the surface of $SrTiO_3$. The air above the solid was flushed with $N_2$ followed by $H_2$ and heated to 500 °C for 1.5 h to remove nitrates and reduce nickel followed by heating in an $O_2$ atmosphere at 130 °C for 30 min adapted from literature procedures [41].

High resolution transmission electron microscopy (HRTEM) and scanning transmission electron microscopy (HRSTEM) images were taken using a JEOL 2500SE 200 kV TEM. To prepare samples, copper grids with a carbon film were dropped into aqueous dispersions of the samples followed by washing with water and air drying.

UV/Vis diffuse reflectance spectra were recorded on dried powders on white Teflon tape using a Thermo Scientific Evolution 220.

For electrochemical measurements, thin films of the catalysts were prepared on a gold foil electrode (1.0 $cm^2$) by drop coating and annealing at 25 °C. A wire was attached to the bare gold back with conductive carbon tape and sealed with adhesive. The electrode was placed into a $N_2$-purged 3-electrode cell with a Pt counter electrode and a saturated calomel reference electrode connected to the cell with a KCl salt bridge. The cell was filled to 50 mL with 0.25 M $Na_2HPO_4/NaH_2PO_4$ buffer solution at pH 7 for overpotential scans and 0.1 M KCl with 10 % (vol) methanol for photocurrent onset measurements, each with a constant stream of $N_2$ above the solution. Cyclic voltammetry scans were taken at 50 mV/s. The system was calibrated using the redox potential of $K_4[Fe(CN)_6]$ at +0.358 V (NHE).

The rate of photochemical hydrogen and oxygen evolution from each catalyst was determined by irradiating 50 mg of NiO-STO dispersed in 50 mL of water. Irradiations were performed in a quartz round bottom flask with a 300 W Xe arc lamp (26.3 mW $cm^{-2}$ at the flask $\lambda = 250-380$ nm), measured with a GaN photodetector. The air-tight irradiation system connects a vacuum pump and a gas chromatograph (Varian 3800) with the sample flask to quantify the amount of gas evolved, using area counts of the peaks and the identity of the gas from the calibrated carrier times. Prior to irradiation, the flask was evacuated down to 5 torr and purged with argon gas. This cycle was repeated until the chromatogram of the atmosphere above the solution indicated that the sample did not contain hydrogen, oxygen, or nitrogen.

Powder XRD scans were conducted with a Scintag XRD, $\lambda = 0.154$ nm at −45 kV and 40 mA with tube slit divergence (2 mm), scatter (4 mm), column scatter (0.5 mm), and receiving (0.2 mm). The FWHM of the peaks were calculated with Themo ARLDMSNT software (version 1.39-1).

**Acknowledgments** Financial support was provided by Research Corporation for Science Advancement (Scialog award), by the National Science Foundation (NSF, Grants 0829142 and 1133099) and by the U.S. 70 Department of Energy under Grant FG02-03ER46057. TKT thanks NSFGRFP for fellowship 2012.

## References

1. G.N. Baum, J. Perez, K.N. Baum, DOE **1**, 1–127 (2009)
2. F.A. Frame, T.K. Townsend, R.L. Chamousis, E.M. Sabio, T. Dittrich, N.D. Browning, F.E. Osterloh, J. Am. Chem. Soc. **133**, 7264–7267 (2011)
3. T.K. Townsend, E.M. Sabio, N.D. Browning, F.E. Osterloh, Energy Environ. Sci. **4**, 4270–4275 (2011)
4. S.U.M. Khan, J. Akikusa, J. Phys. Chem. B **103**, 7184–7189 (1999)
5. N. Beermann, L. Vayssieres, S.E. Lindquist, A. Hagfeldt, J. Electrochem. Soc. **147**, 2456–2461 (2000)
6. A. Duret, M. Gratzel, J. Phys. Chem. B **109**, 17184–17191 (2005)
7. I. Cesar, A. Kay, J.A.G. Martinez, M. Gratzel, J. Am. Chem. Soc. **128**, 4582–4583 (2006)
8. R.K. Hocking, R. Brimblecombe, L.Y. Chang, A. Singh, M.H. Cheah, C. Glover, W.H. Casey, L. Spiccia, Nat. Chem. **3**, 461–466 (2011)
9. N. Sakai, Y. Ebina, K. Takada, T. Sasaki, J. Phys. Chem. B **109**, 9651–9655 (2005)
10. B.A. Pinaud, Z.B. Chen, D.N. Abram, T.F. Jaramillo, J. Phys. Chem. C **115**, 11830–11838 (2011)
11. K. Saito, K. Koga, A. Kudo, Dalton Trans. **40**, 3909–3913 (2011)
12. S.C. Yan, L.J. Wan, Z.S. Li, Z.G. Zou, Chem. Commun. **47**, 5632–5634 (2011)
13. T. Yokoi, J. Sakuma, K. Maeda, K. Domen, T. Tatsumi, J.N. Kondo, Phys. Chem. Chem. Phys. **13**, 2563–2570 (2011)
14. M.S. Wrighton, A.B. Ellis, P.T. Wolczanski, D.L. Morse, H.B. Abrahamson, D.S. Ginley, J. Am. Chem. Soc. **98**, 2774–2779 (1976)

15. K. Domen, S. Naito, M. Soma, T. Onishi, K. Tamaru, J. Chem. Soc., Chem. Commun. 543–544 (1980)
16. F.T. Wagner, G.A. Somorjai, Nature **285**, 559–560 (1980)
17. A. Kumar, P.G. Santangelo, N.S. Lewis, J. Phys. Chem. **96**, 834–842 (1992)
18. K. Sayama, H. Arakawa, J. Photoch, Photobiol. A **77**, 243–247 (1994)
19. K. Sayama, K. Mukasa, R. Abe, Y. Abe, H. Arakawa, Chem. Commun. 2416–2417 (2001)
20. R. Konta, T. Ishii, H. Kato, A. Kudo, J. Phys. Chem. B **108**, 8992–8995 (2004)
21. J.W. Liu, G. Chen, Z.H. Li, Z.G. Zhang, J. Solid State Chem. **179**, 3704–3708 (2006)
22. Y. Sasaki, H. Nemoto, K. Saito, A. Kudo, J. Phys. Chem. C **113**, 17536–17542 (2009)
23. K. Iwashina, A. Kudo, J. Am. Chem. Soc. **133**, 13272–13275 (2011)
24. J.M. Lehn, J.P. Sauvage, R. Ziessel, New J. Chem. **4**, 623–627 (1980)
25. J.M. Lehn, J.P. Sauvage, R. Ziessel, L. Hilaire, Israel J. Chem. **22**, 168–172 (1982)
26. A. Kudo, Y. Miseki, Chem. Soc. Rev. **38**, 253–278 (2009)
27. K. Domen, S. Naito, T. Onishi, K. Tamaru, Chem. Phys. Lett. **92**, 433–434 (1982)
28. P.K. Roy, J. Bera, Mater. Res. Bull. **40**, 599–604 (2005)
29. T. Tsumura, K. Matsuoka, M. Toyoda, J. Mater. Sci. Technol. **26**, 33–38 (2010)
30. C.W. Beier, M.A. Cuevas, R.L. Brutchey, J. Mater. Chem. **20**, 5074–5079 (2010)
31. K. v. Benthem, C. Elsasser, R.H. French, J. Appl. Phys. **90**, 6156–6164 (2001)
32. Y. Mune, H. Ohta, K. Koumoto, T. Mizoguchi, Y. Ikuhara, Appl. Phys. Lett. **91**, 192105 (2007)
33. K. Suttiponparnit, J. Jiang, M. Sahu, S. Suvachittanont, T. Charinpanitkul, P. Biswas, Nanoscale Res. Lett. **6**, 27 (2011)
34. T.K. Townsend, E.M. Sabio, N.D. Browning, F.E. Osterloh, Chemsuschem **4**, 185–190 (2011)
35. E.M. Sabio, R.L. Chamousis, N.D. Browning, F.E. Osterloh, J. Phys. Chem. C **116**, 3161–3170 (2012)
36. A.J. Nozik, Annu. Rev. Phys. Chem. **29**, 189–222 (1978)
37. K. Nakaoka, J. Ueyama, K. Ogura, J. Electroanal. Chem. **571**, 93–99 (2004)
38. M.R. Allen, A. Thibert, E.M. Sabio, N.D. Browning, D.S. Larsen, F.E. Osterloh, Chem. Mater. **22**, 1220–1228 (2010)
39. R. Baba, A. Fujishima, J. Electroanal. Chem. Interfacial Electrochem. **213**, 319–321 (1986)
40. J. Bera, D. Sarkar, J. Electroceram. **11**, 131–137 (2003)
41. K. Domen, A. Kudo, M. Shibata, A. Tanaka, K. Maruya, T. Onishi, J. Chem .Soci.-Chem. Commun. 1706–1707 (1986)

# Chapter 5
# Complete Water Splitting with Multi-Component Catalysts: Proposed Mechanism of Charge Transport in NiOx Loaded $SrTiO_3$ Photocatalyst for Complete Water Splitting

## 5.1 Introduction

Designing a photocatalyst for direct conversion of solar energy to storable chemical fuel may be one of the most important challenges of the century. Stoichiometric water splitting to hydrogen and oxygen gas has been determined to be the cheapest option for renewable hydrogen generation [1]. Photocatalysts that are capable of simultaneously reducing and oxidizing water do not require sacrificial reagents or redox mediators. Wide band gap semiconductors can achieve this as long as their conduction (CB) and valence bands (VB) straddle the water red/ox potentials with a band gap Eg >1.23 eV. The absence of sacrificial reagents creates a true power generating catalyst and the absence of redox mediators eliminates slow aqueous charge diffusion processes. These catalysts often employ co-catalysts for increased separation of the photogenerated electron/hole pairs (i.e. Pt, Pd, Ni, Ru for water reduction and $RuO_2$, $IrO_2$ for water oxidation) [2]. Many of these cocatalyst materials are rare and expensive, and Pt is known to also catalyze the reverse reaction to reduce $O_2$ [3].

Nickel oxide loaded onto strontium titanate (STO) has been demonstrated to act as a complete water splitter from pure water and NaOH solutions [4–7]. The mechanism proposed for this catalyst involves electron tunneling via the conduction band of STO through nickel and into NiOx for water reduction ($STO_{CB} \rightarrow Ni \rightarrow NiO \rightarrow H^+/H_2$). This would leave holes near the valance band of STO for water oxidation ($STO_{VB} \rightarrow H_2O/O_2$). However, this is a doubly combined Schottky barrier where $Ni^0$ acts as a conductor of electrons and is otherwise not reactive with the solution. Considering that NiO is a p-type large band gap semiconductor (3.0–4.5 eV), pairing it with n-type STO forms a p/n junction with Fermi level matching. This implies the existence of a reversed

This chapter appeared as: "T.-K. Townsend, et al., *Nanoscale Strontium Titanate Photocatalysts for Overall Water Splitting*, ACS Nano. **6**, 7420–7426 (2012)."

T. K. Townsend, *Inorganic Metal Oxide Nanocrystal Photocatalysts for Solar Fuel Generation from Water*, Springer Theses, DOI: 10.1007/978-3-319-05242-7_5,

photophysics model where $Ni^0$ accepts electrons from the CB of STO ($STO_{CB} \rightarrow Ni \rightarrow H^+/H_2$) and NiO accepts holes from the VB of STO ($STO_{VB} \rightarrow NiO \rightarrow H_2O/O_2$) [8]. In this way, proton reduction should occur on $Ni^0$ sites and water oxidation on NiO sites. This model is supported by surface photovoltage (SPV) measurements, electrochemistry, photo-labeling experiments in addition to photocatalytic water splitting irradiation tests on Ni-STO, NiOx-STO (where $0 \geq x \geq 1$) and NiO-STO.

## 5.2 Results and Discussion

STO [1] was prepared from a high temperature solid state synthesis from precursors [9]. This material was then modified with nickel and/or nickel oxide at a 3 wt.% NiO:STO loading (Scheme 5.1). Nickel attachment was achieved via three unique methods. First, capped nickel nanoparticles were co-precipitated with STO followed by reductive annealing to form Ni(add)-STO [2]. Second, STO was irradiated in a $Ni^{2+}$ solution to photodeposit nickel to form Ni(h$\nu$)-STO [3]. Third, $Ni(NO_3)_{2(s)}$ was dried on STO, followed by reductive annealing and re-oxidation at 130 °C to form *R*500-*O*130-NiOx-STO [5] and at 500 °C to form *R*500-*O*500-NiO-STO [6]. Finally, nickel was photodeposited onto NiO-STO in pure water to form a three component assembled catalyst, Ni(h$\nu$)-*R*500-*O*500-NiO-STO [7].

Crystallinity and phase of STO were verified with powder XRD (Fig. 5.1). HRTEM images (Fig. 5.2a) of STO displayed a particle size range of 100 nm–1 μm. After NiOx pre-treatment, *R*500-*O*130 NiO-STO showed a combination of dark (high Z-contrast) sites of nickel metal and lighter regions of nickel oxide (Fig. 5.2b). After photodeposition of $Ni^{2+}$, Photodeposited NiO-STO show small (4–6 nm) nickel sites distributed across the STO particles (Fig. 5.2c). Attachment of Ni NP and annealing at 500 °C under $H_2$ pre-treatment formed discrete 5-15 nm nickel nanoparticles covering the surface of STO (Fig. 5.2d). Upon photodeposition of Ni onto R500-O500-NiO-STO, nickel particles (5–10 nm) became larger than in Ni(h$\nu$)-STO whereas NiO particles remained largest (25–50 nm, Fig. 5.2e, f).

STO is known to have an indirect band gap of 3.20 eV (O 2p → Ti 3d $t_{2g}/e_g$) and a direct band gap of 3.75 eV (O 2p → Sr 4d $t_{2g}/e_g$) [10]. NiO is also a large band gap (3.0–4.5 eV) p-type semiconductor with an absorbance tail in the visible region [11]. UV/Vis diffuse reflectance spectra of unmodified STO (white powder, Fig. 5.3a) and NiO-activated STO catalysts (light-dark grey powders) show an optical absorption onset between 3.0 and 3.2 eV and NiOx attachment was accompanied by various visible light absorption due to the color change after NiOx deposition. After oxidation at 130 °C, both Ni(add)-STO and Ni(h$\nu$)-STO increased visible light absorbance below 3.0 eV. In contrast both *R*500-*O*25-NiO-STO and *R*500-*O*130-NiO-STO showed large visible absorption whereas *R*500-*O*500-NiO-STO did not.

SPV scans show photogenerated charge extraction initiating between 3.0 and 3.2 eV for unmodified STO with CPD max = −70 mV (Fig. 5.3b). After

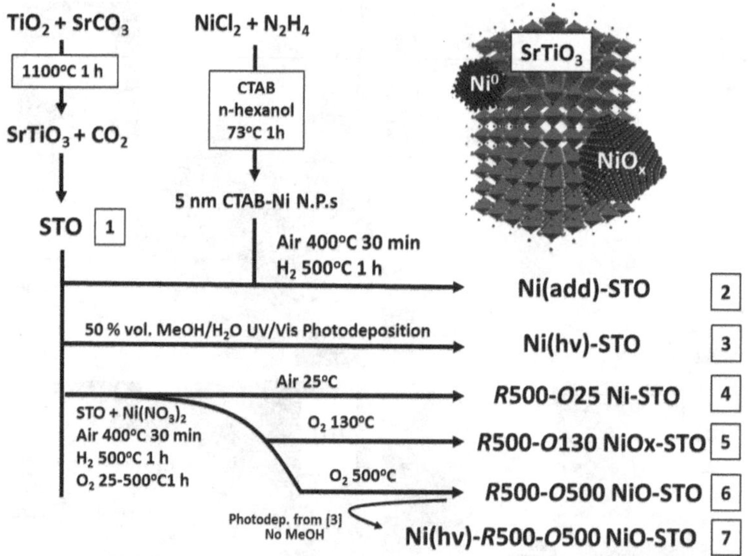


**Scheme 5.1** Synthesis of pure STO [1] and NiOx attached STO [2–7]

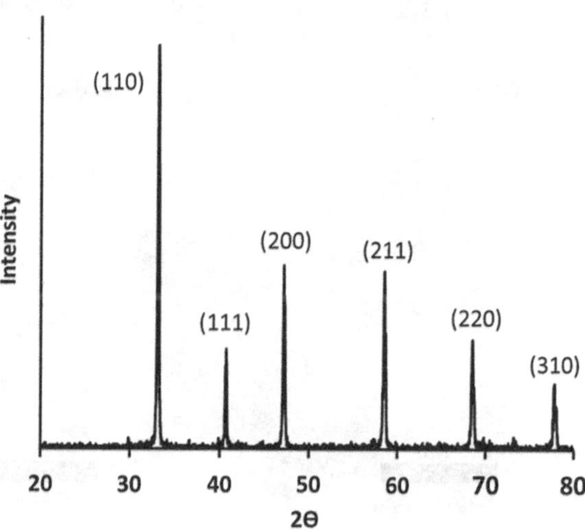


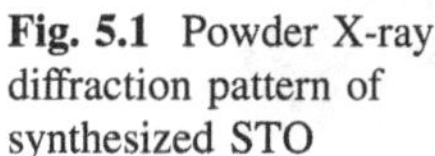
**Fig. 5.1** Powder X-ray diffraction pattern of synthesized STO

attachment of NiOx via redox pretreatments, the ΔCPD of *R*500-*O*130-NiOx-STO increased to −460 mV signifying increased charge separation (Fig. 5.3b) between STO and gold support. A negative −ΔCPD signifies electron injection into the gold substrate due to hole trapping in the catalyst, and this is enhanced with NiOx cocatalyst attachment. In contrast, Ni(hν)-STO showed a positive signal (+370 mV) signifying electron trapping in the material and hole injection into the gold support (Fig. 5.3b). This is also true for Ni(add)-STO, but to a smaller extent (+220 mV).

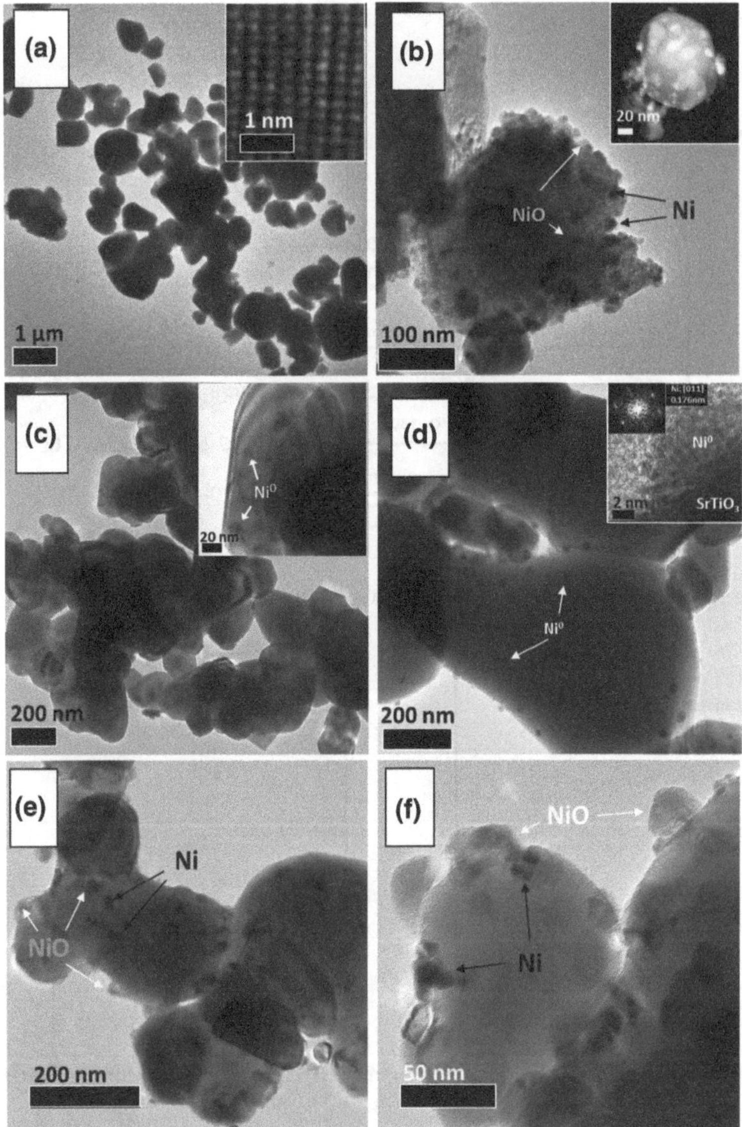


**Fig. 5.2** HRTEM of STO (**a**), HRTEM and STEM of *R*500-*O*130-NiOx-STO (**b**), HRTEM Ni(h*ν*)-STO (**c**), Ni(add)-STO (**d**) and Ni(h*ν*)-*R*500-*O*500-NiO-STO (**e**, **f**)

Since these materials were oxidized at 25 °C, only a fraction of the $Ni^0$ was converted into NiO. $Ni^0$ metal dominates the nickel composition of these cocatalysts. Therefore, we conclude that $Ni^0$ acts as an electron trap and NiO serves as an hole trap for photogenerated electron/hole pairs in STO.

To further confirm the effect of nickel oxidation on charge separation, nickel in a reduced state (*R*500-*O*25 Ni-STO) was heated in series in air from

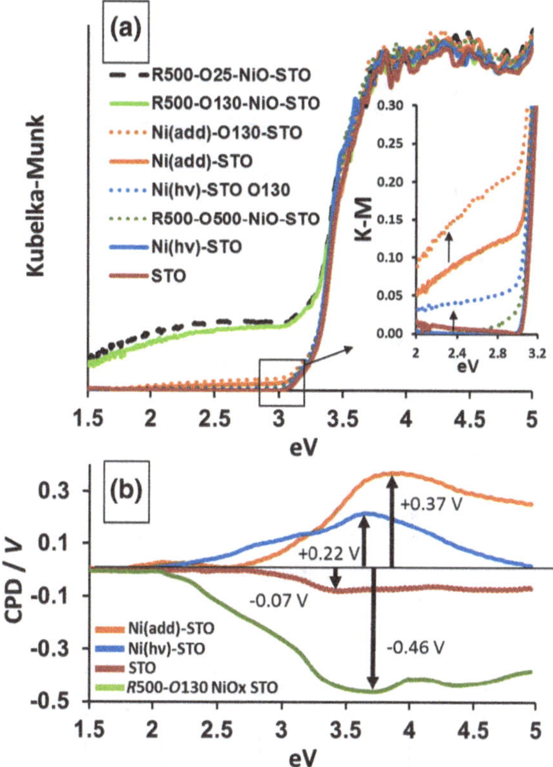


**Fig. 5.3** UV/Vis diffuse reflectance of STO and NiOx modified STO powder (**a**) and light induced contact potential change surface photovoltage measurements on a gold electrode under vacuum ($2 \times 10^{-4}$ mbar) versus excitation energy for STO, Ni-loaded and NiOx-loaded STO with CPD max values (**b**)

25 °C → 50 °C → 100 °C → 130 °C → 500 °C then cooled to 25 °C and tested for photo-generated charge injection (Fig. 5.4a). After heating, nickel metal becomes oxidized to NiOx, and higher temperatures produce higher ratios of NiO:Ni. A reverse in charge transport from electron trapping to hole trapping occurs between 50 and 100 °C in the thin film of NiO-STO and then increases in magnitude with further oxidation (Fig. 5.4a).

This further confirms that Ni metal sites act as electron traps whereas NiOx sites serve as hole traps in NiOx-STO (Fig. 5.4b). The chemical transformation of $Ni^0$ into NiO is also reflected in the morphologies of the materials such that *R*500-*O*25 NiO-STO HRTEM show small (5–10 nm) discrete Ni particles (Fig. 5.4b) which resemble the morphology of Ni(add)-STO (Fig. 5.2d). *R*500-*O*130 NiOx-STO shows a combination of dark Z-contrast Ni particles and light contrast NiOx sites (Fig. 5.4c), whereas *R*500-*O*500 NiO-STO displays large (50–100 nm) NiOx particles without dark Ni sites (Fig. 5.4d), thus verifying the composition change with increased oxidation temperature.

In order to evaluate catalytic activity, NiOx modified STO powders were dispersed in pure water under inert atmosphere and irradiated with a Xe arc lamp. $H_2$ and $O_2$ gases are evolved from the catalyst, bubble above the solution and are quantitated with gas chromatography. Of all the catalysts, *R*500-*O*130-NiO-STO

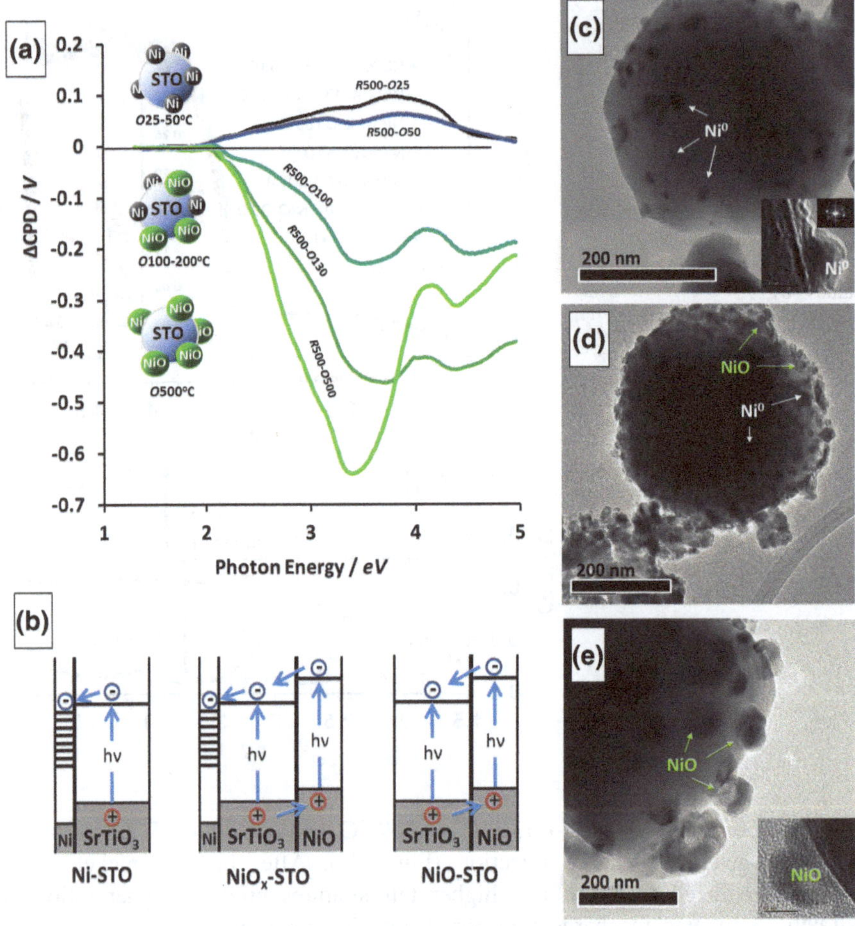


**Fig. 5.4** Light-induced contact potential change ΔCPD of 3 wt.% NiO-STO films on a gold glass electrode under vacuum ($2 \times 10^{-4}$ mbar) versus excitation energy (**a**) after reduction at 500 °C followed by re-oxidation at 25 °C (*black*), 50 °C (*blue*), 100 °C (*cyan*), 130 °C (*dark green*), and 500 °C (*light green*) with proposed photogenerated electron pathways of the three systems (**b**) and HRTEM images of *R*500-*O*25-Ni-STO (**c**), *R*500-*O*130-NiOx-STO (**d**) and *R*500-*O*500-NiO-STO (**e**)

had the highest activity (28 μmol $g^{-1}$ $h^{-1}$, Fig. 5.5a) followed by Ni(hν)-STO (22.5 μmol $g^{-1}$ $h^{-1}$, Fig. 5.5b) which evolved sub-stoichiometric amounts of $O_2$, and Ni(add)-STO (5.3 μmol $g^{-1}$ $h^{-1}$, Fig. 5.5d). The activity of Ni(add)-STO was shown to increase after an oxidation step by heating in $O_2$ at 130 °C for 1 h (Fig. 5.5d). Fully oxidized R500-O500-NiO-STO was not active (Fig. 5.5c), signifying the importance of Ni metal sites. Through a building block approach, NiOx-STO was alternatively prepared by photodepositing Ni metal onto the inactive *R*500-*O*500-NiO-STO to give Ni(hν)-R500-O500-NiO-STO. This process

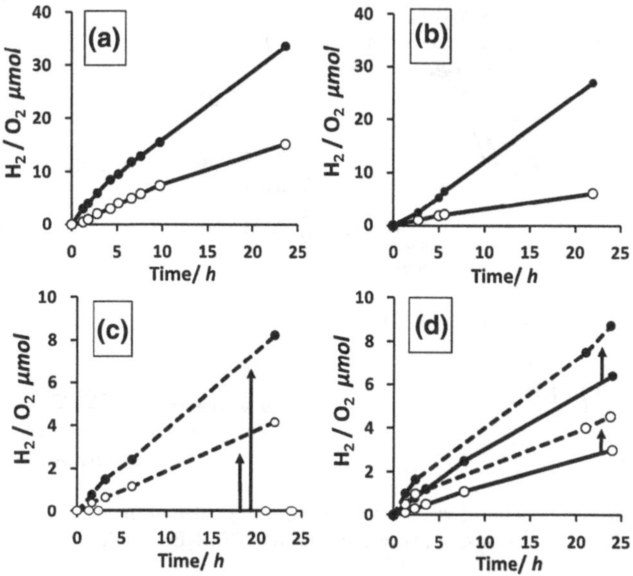


**Fig. 5.5** $H_2$ and $O_2$ evolution from 272 μmol (50 mg) NiOx loaded STO suspended in 50 mL water at pH 7 under full spectrum irradiation with catalyst pre-treatments (**a**) *R*500-*O*130-NiOx-STO, (**b**) Ni(hν)-STO, (**c**) Inactive *R*500-*O*500-NiO-STO (*solid line*) and activated Ni(hν)-*R*500-*O*500-NiO-STO (*dashed line*), (**d**) Ni(add)-STO-*R*500-*O*25 (*solid line*) and Ni(add)-STO-*R*500-*O*130 (*dashed line*)

re-activated the three-component catalyst (Fig. 5.5c) providing further evidence for the necessity of both $Ni^0$ and NiO sites for water reduction and oxidation, respectively.

Dark electrochemical scans from films of the catalysts reveal the overpotentials ($\eta$) for water reduction (Fig. 5.6a) and oxidation (Fig. 5.6b), in reference to theoretical values at pH 7 (−0.413 V vs. NHE $H^+/H_2$ and +0.817 V vs. NHE $H_2O/O_2$). Unmodified STO showed a reduction potential of −0.85 V ($\eta = 0.44$ V) which was shifted to less negative values upon NiOx attachment by +0.05 V for Ni(hν)-STO, by +0.08 V for Ni(add)-STO and by +0.09 V for *R*500-*O*130 NiOx-STO. From this, the proton reduction overpotentials followed the trend of plain NiO (0.467 V) > STO (0.437 V) > Ni(hν)-STO (0.387 V) > Ni(add)-STO (0.357 V) > *R*500-*O*130 NiO-STO (0.347 V).

The water oxidation dark processes show a similar lowering of overpotentials for water oxidation to less positive values compared unmodified STO. Water oxidation potentials shifted from STO (+1.62 V) by −0.02 V for Ni(add)-STO, −0.03 V for Ni(hν)-STO, −0.09 V for *R*500-*O*130 NiOx-STO, and −0.19 V for plain NiO. From this, the overpotentials for water oxidation follow the trend: STO (+0.80 V) > Ni(add)-STO (+0.78 V) > Ni(hν)-STO (+0.77 V) > Plain NiO (+0.61 V). Bubbles formed on the electrode in each case indicating that the positive current resulted from water oxidation instead of other oxidation reactions (i.e. $Ni^0 \rightarrow$

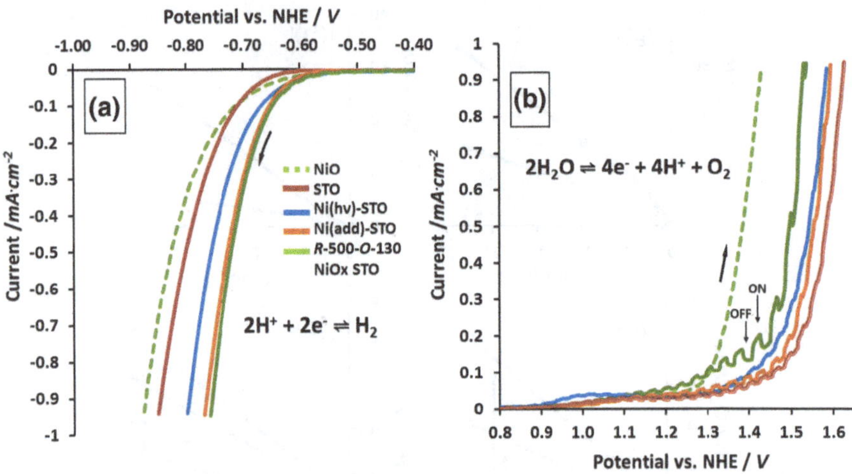


**Fig. 5.6** Photo-electrochemical water reduction (**a**) and oxidation (**b**) with NiOx-STO films on Au substrate, 10 mV $s^{-1}$ scans, at pH = 7 in $Na_2HPO_4/NaH_2PO_4$ buffer

$Ni^{2+} \rightarrow Ni^{3+/4+}$). The low water oxidation overpotential of Plain NiO suggests that it is capable of more efficient hole transport than STO, and that catalysts with a higher ratio of NiO:Ni have a lower barrier for water oxidation to $O_2$. In support of the p/n junction tandem model, light absorption and light-induced charge generation in NiO was observed with diffuse reflectance and SPV (Fig. 5.7). Nickel oxide gives a positive ΔCPD indicating photo-electron trapping and photo-hole injection into the Au substrate. The preferential hole transporting properties of NiO are known and have been used to improve the charge separation in OLED devices [12, 13]. The CPD decreases after 3.75 eV due to slow recombination of photo-generated electron/holes as a consequence of decreasing light intensity (Fig. 5.7). This is also observed in the other NiOx-STO catalysts (Figs. 5.3b, 5.4a).

The light absorbing qualities of NiO cocatalyst are also shown in the UV/Vis diffuse reflectance (Fig. 5.3a) for the NiOx modified STO in addition to the photoelectrochemistry (Fig. 5.8a). For this, catalysts were drop-coated onto an Au electrode in an electrochemical cell and illuminated with chopped light to measure anodic photocurrent. Plain STO showed the highest photocurrent which decreased after NiOx attachment. This is most likely due to UV-light absorption competition with STO. The photo-onset marks the potential where the photocurrent approaches zero and signifies the electron Fermi level ($E_{Fn}$). Attachment of p-type NiO caused a photo-onset shift to more positive potentials due to Fermi level matching with n-type STO. Photo-onset values for the NiOx-STO catalysts with water reduction/oxidation potentials and valence band edges calculated from the absorbance onset indirect band gap are listed in Fig. 5.8b.

In this case, attaching NiOx, depresses the $E_{Fn}$ of the electron to more positive values from unmodified STO (−0.87 V) by +0.20 V for Ni(add)-STO, +0.47 V for R500-O130-NiOx-STO, and +0.58 V for Ni(hν)-STO (Fig. 5.8) which is a known

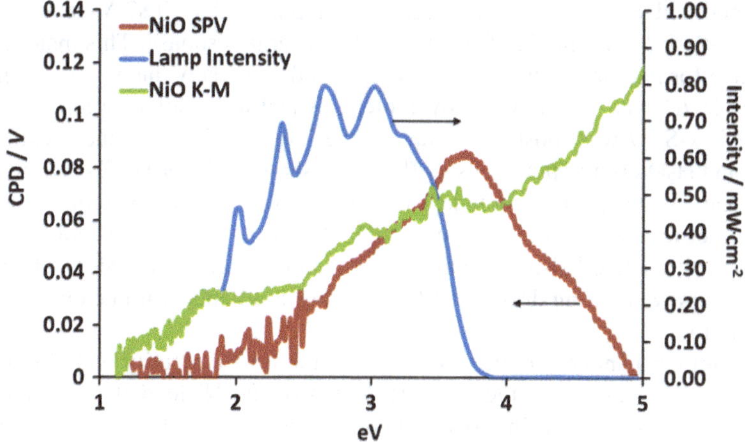


**Fig. 5.7** Light-induced contact potential change from SPV of plain NiO (*left axis*) with NiO light absorbance diffuse reflectance and lamp intensity spectra (*right axis*)

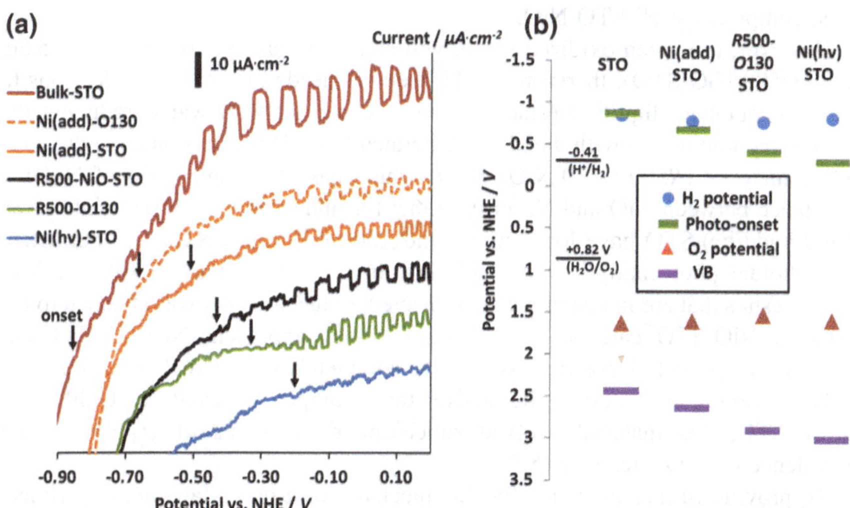


**Fig. 5.8** Photo-current scans of NiOx-STO films on Au substrate, 10 mV $s^{-1}$ scans, at pH = 7 in 0.1 M KCl 10 % vol. methanol (**a**) with corresponding energy diagrams of catalysts (**b**)

effect of cocatalyst addition [14]. Both Ni(add)-STO and R500-O130-NiO-STO contain randomly distributed NiOx islands on the STO; however, Ni(hν)-STO is likely "wired" up for site specific $STO_{CB}$ electron donation to NiOx. This could explain the larger Fermi level depression of n-type STO when paired with p-type NiO. Given that the $E_{Fn}$ level of NiO (+0.67 V) [15] is lower than STO (−0.87 V), Fermi level depression is expected for STO when forming a p/n junction with NiO.

Unmodified STO shows the highest onset potential (−0.87 V) and a small ΔCPD (−0.07 V) originating from a low photocharge response. This material was not active for photocatalytic water splitting and this is likely due to a combination of large $H^+/H_2$ and also $H_2O/O_2$ overpotentials and low photocharge. The *R*500-*O*130-NiOx-STO was most active due to the large ΔCPD combined with lower $H^+/H_2$ and $H_2O/O_2$ overpotentials. Ni(hν)-STO showed higher $H^+/H_2$ activity most likely due to site-specific photodeposition; however, the $O_2$ evolution was sub-stoichiometric due to the improper Ni:NiO ratios. Trace methanol from the photodeposition could also cause this increase in $H_2$ production over $O_2$; however, this was removed during drying at 130 °C and re-irradiation with no change in the $H_2$:$O_2$ ratio.

In a similar manner, Ni(add)-STO showed the lowest activity with the highest $H_2O/O_2$ overpotential and the smallest ΔCPD of the Ni loaded materials. This could likewise stem from improper Ni:NiO ratios since NiO is needed for a low water oxidation overpotential. Upon heating to 130 °C, the activity of this material improves (Fig. 5.5d), which can be attributed to a higher NiO percent which provide hole transport and reduce exciton recombination. Also, nickel metal sites still remain since the material was not fully oxidized and the catalyst operates with three components: Ni-STO-NiO.

However, too much oxidized nickel will create an inactive material (Fig. 5.5c, *R*500-*O*500-NiO-STO), therefore the Ni metal is needed to transport electrons to the semiconductor-liquid interface. There, it can react with water, reducing the level of recombination with the photo-generated hole. This can be seen in the most active material (*R*500-*O*130-NiOx-STO) which has the proper ratio of Ni:NiO. A balance between NiO and Ni must be met for higher efficiency materials given that Ni(add/hν)-STO have low activity and *R*500-*O*500 NiO-STO was not active for colloidal photocatalytic water splitting. In order to isolate Ni from NiO as separate sites that are not chemically connected, reduced nickel was attached to the oxidized NiO-STO catalyst. In a building block approach, Ni-STO-NiO was constructed by photodepositing Ni onto the completely oxidized *R*500-*O*500-NiO-STO in order to generate the active three-component catalyst. Unlike the Ni(hν)-STO, this material evolved stoichiometric amounts of $O_2$ due to the prevalence of NiO sites (Fig. 5.5c).

To provide further evidence for this mechanism, a photo-labeling experiment was conducted to determine the active sites for water reduction/oxidation. Photochemical reduction of platinum (IV) to platinum (0) onto the photo-electron centers of the NiO-STO system was carried out via photodeposition onto the completely oxidized R500-O500-NiO-STO colloid to produce Pt(hν)-R500-O50-NiO-STO. The morphology of this catalyst (Fig. 5.9) showed individual Pt islands of high STEM z-contrast on the surface of STO (site of electron injection into solution), whereas the lower contrast NiO islands remain unreacted. Since Pt did not reductively deposit onto the NiO sites, NiO should not supply electrons to the solution.

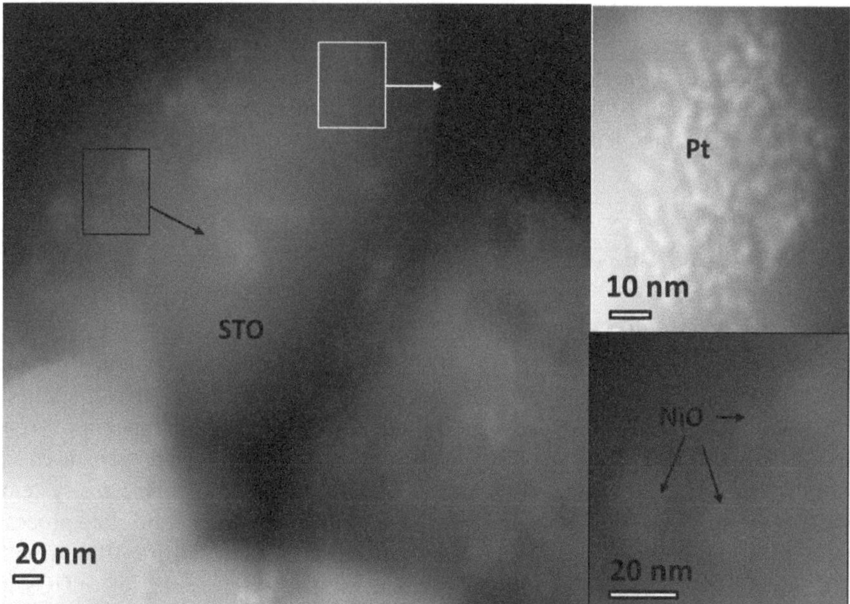


**Fig. 5.9** HRSTEM image of Pt(hν)-*R*500-*O*500-NiO-STO with *insets* showing Pt and NiO islands

## 5.3 Conclusion

NiOx loading onto STO activates the material for photocatalytic water splitting in pure water under UV-light irradiation. The NiOx cocatalyst accepts photogenerated excitons and prolongs their lifetimes by increasing the spatial separation of charge, and the composition of NiOx ($0 \leq x \leq 1$) is crucial for activating the material. Here we show the charge transport pathway through the catalysts and identify the active sites for water reduction (Ni) and water oxidation (NiO). P-type materials like NiO are usually water reducers, not water oxidizers, due to band bending at the water-NiO interface. In this case, however, NiO acts as the water oxidizer. This may be enabled by Ni doping of NiO to become more n-type, or because NiO is nanoscale, and thus band bending is less important than charge transfer kinetics at the NiO/$H_2O$ and NiO/STO interfaces. Catalysts loaded with higher ratios of Ni metal were found to be less active and have a positive ΔCPD value signifying electron trapping on the metal sites. In contrast, catalysts loaded with higher ratios of NiO were less active and showed large negative ΔCPD values which correspond to hole trapping in NiO. Only materials with proper ratios of both Ni and NiO sites were found to be catalytically active in stoichiometric amounts. This was also reflected in the electrochemical scans where active materials had a low $H^+/H_2$ and $H_2O/O_2$ overpotential for water splitting. In order to verify that solitary Ni and NiO sites were the key to an active material, the inactive completely oxidized form of NiO-STO (*R*500-*O*500-NiO-STO) was loaded with Ni metal via photodeposition, re-activating it as a three-component photocatalyst for complete water splitting. Photo-labeling from photoelectrochemical reduction of Pt(IV) to Pt(0) on the STO surface instead of NiO islands further confirms this electronic pathway.

## 5.4 Experimental

Chemicals (KOH 99.9 %, Oxalic acid 99.9 %, $Sr(NO_3)_2$ 99.9 %, P25 $TiO_2$, $Ni(H_2O)_6(NO_3)_2$ 99.9 %, $NiCl_2 6H_2O$ 99.9 %, cetyltrimethylammonium Bromide (CTAB) 99.9 %, *n*-hexanol 99.9 %) were purchased from Fisher Scientific, Pittsburg, PA. $Sr(OH)_2$ was prepared by precipitation at room temperature from $Sr(NO_3)_2$ and KOH in water. They were of reagent quality and were used as received. Water was purified to 18 MΩ cm resistivity using a *Nano-pure* system.

**STO particles:** STO was synthesized via a high temperature solid state process [9] where 0.02 mol, 1.74 g P25 $TiO_2$ was mixed and sonicated for 10 min with 0.026 mol, 5.53 g $Sr(NO_3)_2$ in a 1:1.2 molar ratio to produce 4 g of $SrTiO_3$ in 200 mL. Oxalic acid (0.4 M, 1:1 molar ratio with Sr) was added drop wise to the solution under vigorous stirring. Ammonium hydroxide was added slowly to increase the pH of the solution to 6.5 in order to precipitate strontium oxalate crystals onto the 25 nm $TiO_2$ particle surfaces via heterogeneous nucleation [16]. The resulting precipitate was centrifuged and washed 8 times in 50 mL of water, followed by drying in air at 100 °C.This precursor was calcined in a Thermolyne 79300 Tube Furnace to 1100 °C for 1 h with a heating rate of 10 °C/min. After cooling to room temperature, the white solid was washed twice in 50 mL of 5 M $HNO_3$ to remove excess $SrCO_3$ followed by repeated water washes until the supernatant reached a pH of 7.

**NiO attachment-Reduction/Oxidation:** STO particles (200 mg, 1.10 mmol) were added to an aqueous $Ni(NO_3){\cdot}6H_2O$ solution (0.023 g, 3 wt% loading of NiO:STO) and thoroughly mixed in a sonication bath for 10 min. This solution was dried annealed at 400 °C in air for 30 min to form NiO islands on the surface of $SrTiO_3$. The air above the solid was flushed with $N_2$ followed by $H_2$ and heated to 500 °C for 1.5 h to convert $NO_3^-$ to $N_{2(g)}$ and $H_2O_{(l)}$ and simultaneously reduce and anneal $Ni^{2+}$ to $Ni^0$ onto STO. This was followed by heating under $O_2$ atmosphere at 25, 50, 100, 130, or 500 °C for 30 min (Re-oxidizing the outer layer of the Ni particles to NiOx, $0 \leq x \leq 1$) [17].

**Ni attachment-Photodeposition:** STO particles (50 mg, $2.7 \times 10^{-4}$ mol) were suspended in 50 mL of (0.013 M, 0.195 g) $NiCl_2.6H_2O_{(aq.)}$ solution that contained 50 % methanol by volume. This mixture was irradiated using a Xenon arc lamp for 24 h until the solution changed from white to dark grey [18]. The resulting grey solid was centrifuged and washed 4 times with 50 mL of water followed by drying at room temperature.

**Pt or Ni-Photodeposition on NiO-STO:** Fully oxidized R500-O500-NiO-STO particles containing Nickel islands were prepared by following steps for "NiO attachment-reduction/oxidation" outline above with a 500 °C re-oxidation temperature. This was followed by Ni attachment via photodeposition of $NiCl_2 6H_2O$ (0.013 M) in pure water under 24 h Xe arc lamp then washed four times in 50 mL $H_2O$. No methanol was used here since irradiation in methanol can lead to back conversion of NiO to Ni in NiO-STO. For Pt(hν)-R500-O500-NiO-STO, $H_2PtCl_6$ (1 wt.%Pt) was used in place of $NiCl_2 6H_2O$.

**NiO attachment via Nickel nanoparticle assembly:** Nickel nanoparticles (5–10 nm) were synthesized following a literature method [19]. Two separate microemusions ($NiCl_2$ [0.1 M]/CTAB/n-hexanol) and ($N_4H_5OH$ [3 M]/CTAB/n-hexanol, adjusted to pH 13 with conc. ammonia) were mixed giving a final wt % ratio = 22 % $H_2O$:33 % CTAB:45 % n-hexanol. This solution was heated and stirred at 73 °C for 1 h. STO particles (200 mg, 1.10 mmol) were added to this CTAB/n-hexanol/water stock solution containing ($2 \times 10^{-5}$ mol, 1.1 mg Ni for 3 wt % NiO) nickel nanoparticles. This dispersion was mixed in a sonication bath for 10 min and the STO and Ni NPs were co-precipitated with 100 mL of 1:1 by vol. mixture of methanol: chloroform. The resulting product was washed 4 times with 50 ml of $H_2O$ and dried at 100 °C. This white solid was heated under air at 400 °C for 1 h followed by $H_2$ reduction at 500 °C for 1 h and exposed to air at 25 °C.

Bright field high resolution transmission electron microscopy (HRTEM) images were taken using a JEOL 2500SE 200 kV TEM and HR scanning transmission electron microscopy (HRTEM) were taken with a JEOL 2100F 200 kV STEM. Copper grids with a carbon film were dropped into aqueous dispersions of the samples followed by washing with water and air drying. UV/Vis diffuse reflectance spectra were taken as dried powders on white Teflon tape using a Thermo Scientific Evolution 220.

For electrochemical measurements, thin films of the catalysts were prepared on a gold foil electrode (1.0 $cm^2$) by drop coating and annealing at 25 °C. A wire was attached to the bare gold back with conductive carbon tape and sealed with adhesive. The electrode was placed into a $N_2$-purged 3-electrode cell with a Pt counter electrode and a saturated calomel reference electrode connected to the cell with a KCl salt bridge. The cell was filled to 50 mL with 0.25 M $Na_2HPO_4$/ $NaH_2PO_4$ buffer solution at pH 7 with a constant stream of $N_2$ above the solution. Dark cyclic voltammetry scans were taken at 50 mV/s. The system was calibrated using the redox potential of $K4[Fe(CN)_6]$ at +0.358 V (NHE). For light measurements, chopped light was introduced using a 300 W Xe arc lamp.

Surface photovoltage (SPV) measurements were conducted under vacuum ($2 \times 10^{-4}$ mBar) on $SrTiO_3$ films on gold substrates. A gold Kelvin probe (Delta PHI Besocke) served as the reference electrode. Samples were illuminated with monochromatic light from a 150 W Xe lamp filtered through a Oriel Cornerstone 130 monochromator (1–10 mW $cm^{-2}$) and CPD spectra were corrected for drift effects by subtracting a dark scan.

The rate of photochemical hydrogen and oxygen evolution from each catalyst was determined by irradiating 50 mg (0.27 mmol) of photocatalyst dispersed in 50 mL of Nano-pure water. Irradiations were performed in a quartz round bottom flask with a 300 W Xe arc lamp (26.3 mW $cm^{-2}$ at the flask $\lambda = 250-380$ nm), measured with a GaN photodetector. The air-tight irradiation system connects a vacuum pump and a gas chromatograph (Varian 3800) with the sample flask to quantify the amount of gas evolved, using area counts of the peaks and the identity of the gas from the calibrated carrier times. Prior to irradiation, the flask was evacuated down to 5 torr and purged with argon gas. This cycle was repeated until the chromatogram of the atmosphere above the solution indicated that the sample did not contain hydrogen, oxygen, or nitrogen.

Powder XRD scans were conducted with a Scintag XRD, $\lambda = 0.154$ nm at −45 kV and 40 mA with tube slit divergence (2 mm), scatter (4 mm), column scatter (0.5 mm), and receiving (0.2 mm).

**Acknowledgments** Financial support was provided by Research Corporation for Science Advancement (Scialog award), by the National Science Foundation (NSF, Grants 0829142 and 1133099) and by the U.S. 70 Department of Energy under Grant FG02-03ER46057.TKT thanks NSFGRFP for fellowship 2012.

## References

1. B.D. James, G.N. Baum, J. Perez, K.N. Baum, DOE Contract Number: GS-10F-009 J DOE Technical Monitor: David Peterson **1**, 1–128 (2009)
2. F.E. Osterloh, Chem. Mater. **20**, 35–54 (2007)
3. H. Arakawa, K. Sayama, Catal. Surv. Jpn. **4**, 75–80 (2000)
4. K. Domen, A. Kudo, M. Shibata, A. Tanaka, K. Maruya, T. Onishi, J. Chem. Soc. Chem. Commun. **23**, 1706–1707 (1986)
5. K. Domen, A. Kudo, T. Onishi, N. Kosugi, H. Kuroda, J. Phys. Chem. **90**, 292–295 (1986)
6. K. Domen, S. Naito, T. Onishi, K. Tamaru, Chem. Phys. Lett. **92**, 433–434 (1982)
7. K. Domen, S. Naito, M. Soma, T. Onishi, K. Tamaru, J. Chem. Soc. Chem. Commun. **12**, 543–544 (1980)
8. R. Baba, A. Fujishima, J. Electroanal. Chem. Interfacial Electrochem. **213**, 319–321 (1986)
9. P.K. Roy, J. Bera, Mater. Res. Bull. **40**, 599–604 (2005)
10. K. v. Benthem, C. Elsasser, R.H. French, Journal of Applied Physics, **90**, 6156–6164 (2001)
11. G. Boschloo, A. Hagfeldt, J. Phys. Chem. B **105**, 3039–3044 (2001)
12. Y. Wang, Q. Niu, C. Hu, W. Wang, M. He, Y. Zhang, S. Li, L. Zhao, X. Wang, J. Xu, Q. Zhu, S. Chen, Opt. Lett. **36**, 1521–1523 (2011)
13. M.D. Irwin, D.B. Buchholz, A.W. Hains, R.P.H. Chang, T.J. Marks, Proc. Natl. Acad. Sci. **105**, 2783–2787 (2008)
14. T.K. Townsend, E.M. Sabio, N.D. Browning, F.E. Osterloh, ChemSusChem **4**, 185–190 (2011)
15. K. Nakaoka, J. Ueyama, K. Ogura, J. Electroanal. Chem. **571**, 93–99 (2004)
16. J. Bera, D. Sarkar, J. Electroceram. **11**, 131–137 (2003)
17. K. Domen, A. Kudo, T. Onishi, J. Catal. **102**, 92–98 (1986)
18. J.M. Lehn, J.P. Sauvage, R. Ziessel, New J. Chem. **4**, 623–627 (1980)
19. D.-H. Chen, S.-H. Wu, Chem. Mater. **12**, 1354–1360 (2000)

# Curriculum Vitae of Troy K. Townsend

## Education

| | |
|---|---|
| **US Naval Research Laboratory** | **Washington, D.C.** |
| *Postdoctoral Fellow in Materials Chemistry* | 2012–present |
| **University of California, Davis** | **Davis, CA** |
| *PhD in Inorganic Chemistry* GPA: **3.98** | 2008–2012 |
| **St. Mary's College of Maryland** | **St. Mary's City, MD** |
| *Bachelor of Arts in Chemistry and Biology GPA*: **3.81** | 2003–2007 |

## Graduate Classes

- CHE 201 Symmetry and Group Theory
- CHE 226 Trans Metal Chem
- CHE 295 Careers in Chemistry
- 2A, 2B, 2C Teaching Assistant
- CHE 205 Spectroscopy
- CHE 228B Main Group Chem
- EMS 230 Electron Microscopy
- EMS 230L Elec Microscopy Lab
- CHE 222 Chem of Nanoparticles
- EMS 289A Electronic Materials
- EMS 232 Advanced T E M
- EMS 232L Advanced T E M Lab
- TTP 289B Other Side of the Meter

T. K. Townsend, *Inorganic Metal Oxide Nanocrystal Photocatalysts for Solar Fuel Generation from Water*, Springer Theses, DOI: 10.1007/978-3-319-05242-7,

## Work and Leadership Experience

**Research and Collaboration** **NRL, UC Davis, 2008–present**

- Worked with PI and colleagues to complete research projects, including writing publications, grant proposals and patent applications

**Outreach and Volunteering**

- Advised undergraduate research summer students on laboratory projects
- Chemistry demonstrations for children at chem club booth (methanol cannon, etc.)

**Article Reviewer** **NRL, 2012–present**

- Peer reviewed 6+ journal articles for Energy and Environmental Science

**General Chemistry Lab and Lecture Teaching Assistant** **UC Davis, NSF 2008–2011**

- Led undergraduate class in all three sections of general chemistry laboratory through prescribed experiments and graded lab reports for 8 sections of 25 students
- Taught weekly lecture for extra-help sessions outside of official lecture including discussion of homework problems and explanations of lecture material. Teacher evaluations (4.8/5.0)
- Graded term exams and lecture homework assignments

**High School Chemistry Teacher, Soccer/Wresting Coach** **The Kiski School, 2007–2008**

- Designed and taught sophomore chemistry class for five sections of students including an honors class, assigned and graded home works and exams, served as Dorm Master of boy's boarding hall, assistant wresting and soccer coach, dinner table head

**Undergraduate Ecology Research Student** **Univ. of Maryland and NSF, 2003–2005**

- Led team of scientists to study Schistosomiasis infection in Lake Malawi, Africa, duties included scuba diving vector snail transect sampling, predator fish transects, data analysis and statistics.

## Nanoscience: Skills and Interests

**Nanocrystal inks for spray-on electronics**: Spin and spray-coat of metallic and semiconducting nanocrystal solutions for photovoltaics, light emitting diodes, and other optoelectronic devices.
**Solar Cells and Modules**: Spray-coat, Spin-coat photovoltaic devices fabrication from nanocrystals solutions and AM 1.5 illuminated I-V curve measurements.
**Photocatalysts**: Design, synthesize and test nanomaterials for water splitting reactions and quantify $H_2$ and $O_2$ evolution from water.
**Air Sensitive Syntheses (Schlenk) and ligand exchange**: Nanocrystal Materials Synthesized and Characterized: $K_4Nb_6O_{17}$, $IrO_2$, $Fe_2O_3$, $WO_3$, $SrTiO_3$, NiO, Ni, Au, Pt, Ag, Ag wires, CdTe, CdSe, ITO, $TiO_2$, ZnO….
**Certifications and Training**: Clean room training, Optical and Dektak Profilometry, Edwards thermal metal evaporator, XPS, High Resolution Transmission Electron Microscopy (HRTEM) + HRSTEM + EDS + EELS + SEM + Digital Micrograph, Powder XRD + Scherrer analysis, Photo-electrochemistry (three-electrode systems), UVVis and Diffuse Reflectance + Kubelka Munk Calculations, FTIR + GC + Pellet Press + Excel + Website Design/Maintenance, Surface Photovoltage (SPV Kelvin Force Probe Measurements), Hazardous Waste Management, Lab Safety and Information Security Training.

## Awards and Honors

**Springer Thesis Award Book Publication** PhD dissertation made available in hard copy 2013 €500
**Press Release Nanowerk Highlight**: Making inorganic solar cells with an airbrush spray. Nanowerk.com Sept 25, 2013.
**National Research Council Fellowship** (NRC) for proposal on spray-on solar cells, 2012–present \$75,000/yr × 3
**NSF-Graduate Research Fellowship** (NSFGRFP), for proposal on dye-sensitized solar cells, 2010–2013 \$30,000/yr × 3
**ICAM-I2CAM Travel Award** June 2011 for ICH2P Conference, \$2,500
**UC Davis Travel Award** June 2011 for ICH2P Conference, \$500
**Outstanding Chem 2 Series TA Award,** UC Davis 2010, \$200
**Bradford Borge Fellowship** for outstanding chemistry graduates, UC Davis 2008
**Block Grant,** free university tuition with TA-ship in Chemistry, UC Davis 2008
**Biology Department Award,** Excellence in African schistosomiasis research, 2007
**ACS College Chemistry Achievement Award** from the Chemical Society of Washington and The American Chemical Society, 2006
**Beta Beta Beta** Biological Honors Society 2007
**NSF Grant** (University of Maryland Research Grant, 2003), \$5000
**Dean's List** for Academic Excellence, SMCM, 2003–2007

**Senior Prefect Position** (School Class President), 2003
**Bishop's Prize, 2003,** awarded to high school senior who offered the most to the community of St. James as a whole.
**Alumni Prize**, 2003, awarded to the high school senior who has interacted with and aided the Alumni Council to the greatest extent.
**Halloway Scholarship**, 2002, awarded to rising senior gentleman and lady who displayed excellence as a scholar, athlete, and leader, $5000.

## Publications

(1) **Assembly of Core-Shell Structures for Photocatalytic Hydrogen Evolution from Aqueous Methanol**. Han Zhou, Erwin M. Sabio, Troy K. Townsend, Tongxiang Fan, Di Zhang, Frank E. Osterloh, Chem. Mater., 2010, 22, 3362–3368.
(2) **Improved Niobate Nanoscroll Photocatalysts for Partial Water Splitting.** Troy K. Townsend, Erwin M. Sabio, Nigel D. Browning, and Frank E. Osterloh, ChemSusChem, 2011, 4, 185–190.
(3) **Photocatalytic Water Oxidation with Nonsensitized $IrO_2$Nanocrystals under Visible and UV Light. F**. Andrew Frame, Troy K. Townsend, Rachel L. Chamousis, Erwin M. Sabio, Th. Dittrich, Nigel D. Browning, and Frank E. Osterloh, Journal of the American Chemical Society, 2011, 133, 7264–7267.
(4) **Photocatalytic Water Oxidation with Suspended alpha-$Fe_2O_3$Particles-Effects of Nanoscaling**. Troy K. Townsend, Erwin M. Sabio, Nigel D. Browning and Frank E. Osterloh, Energy and Environmental Science, 2011, 4, 4270–4275.
(5) **Quantum Confinement Controlled Photocatalytic Water Splitting by Suspended CdSe Nanocrystals**. Michael A. Homes, Troy K. Townsend, and Frank E. Osterloh, Chemical Communications, 2011, 48, 371–373.
(6) **Single-Crystal Tungsten Oxide Nanosheets: Photochemical Water Oxidation in the Quantum Confinement Regime**. Mollie Waller, Troy K. Townsend, Jing Zhao, Erwin M. Sabio, Rachel Chamousis, Nigel D. Browning, and Frank E. Osterloh, Chemistry of Materials, 2012, 24, 698–704.
(7) **Nanoscale Strontium Titanate Photocatalysts for Overall Water Splitting**. Troy K. Townsend, Nigel D. Browning, and Frank E. Osterloh, ACS Nano, 2012, 6, 7420–7426.
(8) **Overall Photocatalytic Water Splitting with NiOx-$SrTiO_3$—A Revised Mechanism.** Troy K. Townsend, Nigel D. Browning, and Frank E. Osterloh, Energy and Environmental Science, 2012, 5, 9543–9550.
(9) **Inorganic Photovoltaic Devices Fabricated Using Nanocrystal Spray Deposition**. Ed Foos, Woojun Yoon, Joe Tischler, and Troy K. Townsend, ACS Applied Materials and Interfaces, 2013, 18, 8828–8832.
(10) **Sintered CdTe Nanocrystal Thin-films: Determination of Optical Constants and Application in Novel Inverted Heterojunction Solar Cells**. Woojun Yoon, Troy K. Townsend, Matthew P. Lumb, Joseph G. Tischler, and Edward E. Foos, Submitted to IEEE.

**(11) Solution Spray Deposition of Nanocrystals for Schottky Solar Cells in Air**. Troy K. Townsend, Woojun Yoon, Joe Tischler, Ed Foos, Manuscript writing stage.

## Professional Meeting Presentations

**(1) ACS Meeting San Francisco**, Inorganic 94, March 21–25, 2010. Title: Effects of simultaneous platinum and iridium dioxide deposition in asymmetric $KNb_6O_{17}$ nanoscrolls on photocatalytic $H_2$ evolution from water. Troy Townsend, Erwin Sabio, Frank E. Osterloh.
**(2) SPIE Meeting San Diego**, 2010 August 1–5, Solar Hydrogen and Nanotechnology V, edited by Hicham Idriss, Heli Wang, Proc. of SPIE Vol. 7770, 77700I • © 2010 SPIE • CCC code: 0277-786X/10/$18 • doi: 10.1117/12. 860063 Proc. of SPIE Vol. 7770 77700I-1. Title: Effects of simultaneous platinum and iridium dioxide deposition in asymmetric $KNb_6O_{17}$ nanoscrolls on photocatalytic $H_2$ evolution from water. Troy Townsend, Erwin Sabio, Frank E. Osterloh.
**(3) ICH2P International Conference on Hydrogen Production, CPERI, Thessaloniki, Greece**, 2011 June 19–22, Photocatalytic Processes, Chairman: K.S. Triantafyllidis, 138PHO. Title: Non-supported alpha-$Fe_2O_3$ nanocrystals as photocatalysts for water oxidation. Troy Townsend, Frank E. Osterloh, E.M Sabio, N. Browning.
**(4) Materials Research Society (MRS) Meeting, Boston, MA**, 2011, Nov 28–Dec 2. Symposium E5 (Co-chair): Advanced Materials for Solar-Fuel Generation. Title: Photocatalytic Water Oxidation with Suspended Alpha-$Fe_2O_3$ Particles—Effects of Nanoscaling.
**(5) Materials Research Society (MRS) Meeting, Boston, MA**, 2013, Dec 1–6. Symposium O: Solution Processing of Inorganic and Hybrid Materials for Electronics and Photonics. Title: Spray Processing of PV Devices.
**(6) Materials Research Society (MRS) Meeting, San Francisco, CA**, 2013, April 1–5. Symposium M: Solution Synthesis of Inorganic Functional Materials—Films, Nanoparticles, and Nanocomposites. Abstract Submitted.

Zeitfracht Medien GmbH
Ferdinand-Jühlke-Straße 7
99095 Erfurt, Deutschland
produktsicherheit@kolibri360.de